A new Silurian (Llandovery, Telychian) sponge assemblage from Gotland, Sweden

by

Freek Rhebergen and Joseph P. Botting

Acknowledgement

Financial support for the publication of this issue of
Fossils and Strata was provided by the Lethaia Foundation

Contents

Introduction 1
Institutional abbreviations 3
Geological setting 3
Material, methods, terminology 6
Systematic palaeontology 8
   Phylum Porifera Grant, 1836 8
   Class Demospongea Sollas, 1875 8
   Order Orchocladina Rauff, 1895 8
   Family Anthaspidellida Miller, 1889 8
      Genus *Archaeoscyphia* Hinde, 1889 8
         Archaeoscyphia minganensis Billings, 1859 8
         Archaeoscyphia annulata Rigby, 1973 12
         Archaeoscyphia gislei de Freitas, 1989 14
         Archaeoscyphia rectilinearis de Freitas, 1989 16
         Archaeoscyphia aulocopiformis de Freitas, 1989 18
         Archaeoscyphia scalaria de Freitas, 1989 19
         Archaeoscyphia attenuata de Freitas, 1989 21
      Genus *Finksella* Rigby & Dixon, 1979 22
         Finksella turbinata Rigby & Dixon, 1979 22
      Genus *Calycocoelia* Bassler, 1927 24
         Calycocoelia typicalis Bassler, 1927 24
      Genus *Climacospongia* Hinde, 1884 25
         Climacospongia undulata de Freitas, 1991 26
      Genus *Multistella* Finks, 1960 28
         Multistella leipnitzae n. sp. 29
      Genus *Somersetella* Rigby & Dixon, 1979 32
         Somersetella amplia Rigby & Chatterton, 1989 32
         Somersetella digitata Rigby & Dixon, 1979 34
   Family Streptosolenidae Johns, 1994 35
      Genus *Postperissocoelia* n. gen. 35
         Postperissocoelia gnisvardensis n. sp. 36
   Family Chiastoclonellidae Rauff, 1895 38
      Genus *Chiastoclonella* Rauff, 1895 39
         Chiastoclonella sp. Rauff, 1895 39
   Subclass Tetractinomorpha Lévi, 1953 39
   Order Streptosclerophorida Dendy, 1924 39
   Suborder Eutaxicladina Rauff, 1894 39
   Family Hindiidae Rauff, 1893 39
      Genus *Hindia* Duncan, 1879 40
         Hindia sphaeroidalis Duncan, 1879 40
   Suborder Sphaerocladina Schrammen, 1910 42
   Family Astylospongiidae Zittel, 1877 42
      Genus *Caryoconus* Rhebergen & van Kempen, 2002 42
         Caryoconus gothlandicus (Schlüter, 1884) 42
      Genus *Lindstroemispongia* n. gen. 45
         Lindstroemispongia cylindrata n. sp. 47
   Order Spirosclerophorida Reid, 1963 48
   Suborder Rhizomorina Zittel, 1878a 48

Family Haplistiidae de Laubenfels, 1955 48
   Genus *Haplistion* Young & Young, 1877 48
      Haplistion minutum Rigby & Dixon, 1979 50
      Haplistion cylindricum Rigby & Dixon, 1979 52
      Haplistion toftanum n. sp. 54
      Haplistion sp. Young & Young, 1877 57
   Genus *Warrigalia* Rigby & Webby, 1988 59
      Warrigalia robusta Rigby & Webby, 1988 59
      Rhizomorina indet. 62
Subclass Tetractininellida? Marshall, 1876 62
Order Hadromerida? Topsent, 1898 63
Family uncertain 63
   Genus *Opetionella* von Zittel, 1878 63
      Opetionella incompta n. sp. 63
Class Hexactinellida Schmidt, 1870 65
Order Lyssacinosida? von Zittel, 1877 65
Family uncertain 65
   Genus *Urphaenomenospongia* n. gen. 65
      Urphaenomenospongia euplectelloides n. sp. 66
Lyssacinosida? indet. 69
Order and family uncertain 71
      Hexactinellid indet. A 71
      Hexactinellid indet. B 71
      Spiculite 72
Class Stromatoporoidea Stearn, Webby, Nestor & Stock, 1999 74
Order Labechiida Kühn, 1927 74
Family Stylostromatidae Webby, 1993 74
   Genus *Pachystylostroma* Nestor, 1964 74
      Pachystylostroma sp. 74
Order Clathrodictyida Bogoyavlenskaya, 1969 76
Family Clathrodictyidae Kühn, 1939 76
Genus *Clathrodictyon* Nicholson & Murie, 1878 76
      Clathrodictyon sp. 76
Order Stromatoporida Stearn, 1980 76
Family Syringostromellidae Stearn, 1980 76
   Genus *Syringostromella* Nestor, 1966 76
      Syringostromella sp. 76
Order Amphiporida Rukhin, 1938 76
Family Amphiporidae Rukhin, 1938 76
Genus *Amphipora* Schulz, 1883 76
      Amphipora sp. 76
Palaeobiogeography and community evolution 77
Taphonomy 80
Palaeoecology and palaeoenvironmental interpretation 80
Conclusions 83
Acknowledgements 83
References 83

A new Silurian (Llandovery, Telychian) sponge assemblage from Gotland, Sweden

FREEK RHEBERGEN AND JOSEPH P. BOTTING

LETHAIA

Rhebergen F. & Botting J. P. 2014: A new Silurian (Llandovery, Telychian) sponge assemblage from Gotland, Sweden. *Fossils and Strata*, Vol. 60, pp. 1–87.

A diverse sponge fauna is reported from the Late Llandovery (Telychian) of Gotland (Sweden) and is the first Silurian assemblage known from Baltica. The fossils have been collected from pebble accumulations along a restricted section of the western coast, but originate from the unexposed 'Red Layer', a poorly known formation underlying the Lower Visby Formation. The age of the assemblage has been determined through co-occurring acritarchs. The assemblage comprises 29 species in 20 genera, 18 species of which have not previously been recorded from Baltica, together with several taxa in open nomenclature. Isolated hexactinellid spicules are common, and these sponges are also represented by rare, partially articulated skeletons. The fauna includes three new genera and six new species: *Multistella leipnitzae* n. sp., *Postperissocoelia gnisvardensis* n. gen. et n. sp., *Lindstroemispongia cylindrata* n. gen. et n. sp., *Haplistion toftanum* n. sp., *Opetionella incompta* n. sp. and *Urphaenomenospongia euplectelloides* n. gen. et n. sp. The endemic *Caryoconus gothlandicus* (Schlüter, 1884) comprises ca. 50 per cent of the assemblage. Early Palaeozoic Rhizomorina are here recorded from Baltica for the first time. The Anthaspidellidae and Rhizomorina show strong similarities with species from younger Silurian strata in Arctic Canada. The named non-lithistid taxa show advanced skeletal architecture more characteristic of Mesozoic sponges. The palaeobiogeography of these faunas suggests that lithistid sponges were strongly affected by the end-Ordovician extinctions, but a variety of Llandoverian taxa recovered to become very widespread and combined with endemic faunas developing within each region. Non-lithistid taxa, in contrast, appear to have undergone important evolutionary development during the Late Ordovician – Early Silurian interval, with at least some modern groups possibly originating in relatively inshore habitats at this time. □ *Gotland, Hexactinellida, Lower Silurian, Orchocladina, palaeobiogeography, palaeoecology, Porifera, Rhizomorina.*

Freek Rhebergen [freek.rhebergen@planet.nl], and Joseph P. Botting [acutipuerilis@yahoo. co.uk], Slenerbrink 178 7812 HJ Emmen, the Netherlands; State Key Laboratory of Palaeobiology and Stratigraphy, Nanjing Institute of Geology and Palaeontology, 39 East Beijing Road, Nanjing, 210008, China;
manuscript received on 16/03/2012; manuscript accepted on 10/10/2013.

'The field being comparatively new and exceedingly difficult, we expect our work to prove faulty in some respects; yet, whatever errors we may have fallen into, we will find consolation in the conviction that we have done the best we could under the circumstances.'

(Ulrich 1890, p. 257)

Introduction

Lower Silurian sponges are globally rare, and Silurian lithistid-dominated faunas are best known from Arctic Canada. A diverse assemblage of silicified sponges, including orchocladines, rhizomorines, stromatoporoids, hexactinellids and non-lithistid demosponges, has been collected from scree accumulations in a restricted area along the western coast of the Island of Gotland, Sweden (Fig. 1, topographical map). The Silurian age (Llandoverian, Tely-chian) of the fauna has been determined through associated acritarchs, brachiopods and tabulates. The assemblage is important in several respects, especially as Early Silurian sponges are globally rare (Muir *et al.* in press), and this assemblage fills one of the many gaps in our knowledge. This is the first known assemblage of Silurian sponges from Baltica and, aside from isolated spicules, includes the first rhizoclonid sponges recorded from the region. The assemblage is very distinct from, and far more diverse than, the assemblage of erratic sponges of Ordovician age that also occurs among the same pebble accumulations on Gotland. These assemblages are compared in detail in a subsequent chapter.

Silurian sponge faunas are generally scarce worldwide, being limited both in species and numbers of specimens when compared with Ordovician assemblages. Rigby & Chatterton (1989), in their description of a Silurian sponge fauna from Baillie-

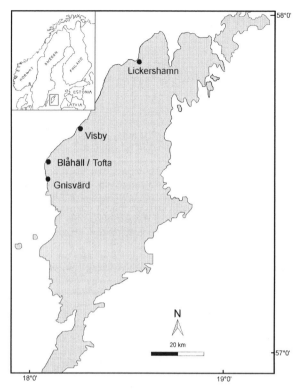

Fig. 1. Map of Gotland (Sweden). About 98 per cent of the collected Telychian sponge specimens have been collected from the Gnisvärd/Blåhäll area.

Hamilton and Cornwallis Islands, Arctic Canada, gave an overview of Silurian sponges known from the literature up to that time, and this is revised and extended by Muir *et al.* (in press). Rigby & Chatterton (1989) summarized about fifty publications, most of them concerning sponges from North America. The only European sponges were those discussed by Rauff (1894); these include two hexactinellids from the UK: *Oncosella* Rauff (1894) and *Amphispongia* Salter (1861), and Ordovician demosponges from northern Europe. Rauff considered *Aulocopium, Astylospongia, Hindia* and *Caryospongia* to be of 'Silurian age'. However, this reflects a widespread stratigraphic confusion, because in older (mainly German) European literature, the Ordovician System was named '(Lower-) Silurian' long after the Ordovician had been erected (Lindström 1888a; Rauff 1893, 1894, 1895; Stolley 1900, 1929). As a consequence, even in recent literature (Finks 2003), Ordovician sponges from Baltica, including Gotland, have erroneously been conceived as of 'Silurian' age.

Palaeozoic sponges from the Swedish mainland are exceptionally rare. Wiman (1907) recorded two erratic specimens, but recent re-investigation in 2004 by FR has shown that Wiman's identification of the specimens as sponges is questionable. The only other record is of three unpublished specimens of *Hindia sphaeroidalis* from the Ordovician Boda-limestone in the Siljan-area (Sweden), which were recognized by FR in the Leipnitz Collection (Uelzen, Germany) and in the collections of the Swedish Museum of Natural History. There are no reports of sponges from Öland, from either bedrock or erratics (Christina Franzén, personal communication 2004). The occurrence of thousands of sponge specimens on Gotland is therefore remarkable.

Sponges from Gotland have been poorly recorded, compared with the numerous papers on erratic sponges collected in Germany and the Netherlands. Lindström (1885, 1888b) listed erratic sponges from Gotland, assigning some of them to his Zone b and Zone c, and relating these to 'the Red Layers'. Rauff (1894) apparently visited the collections of the Swedish Museum of Natural History, as he described and figured a number of specimens. However, there is no indication that he considered any of those sponges to be of 'Upper Silurian' age. Rhebergen & van Kempen (2002) summarized a discussion between Lindström, Schmidt (1891) and Stolley (Stolley 1900, 1929) on the age and lithostratigraphical origin of the sponges. Lindström probably intended to describe at least part of the sponge assemblage, because he had drawings made by Lillj-eval, but did not complete the work (Rhebergen 2007). Schlüter (1884) reported on a remarkable sponge specimen, *Astylospongia gothlandica*, which has been figured by Rauff (1894) as an aberrant form of the Ordovician species *Caryspongia diadema*. Retrospectively, this specimen appears to have been the first specimen of the Silurian assemblage to be figured, and it is now designated as the holotype of *Caryoconus gothlandicus* (Schlüter, 1884).

Van Kempen (1983) described another aberrant erratic specimen from Gnisvärd (Gotland) and followed Rauff's determination. Additional specimens of identical or similar aberrant forms, collected by Mrs. Heilwig Leipnitz, Uelzen (Germany), led to further investigation of the collections of the Swedish Museum of Natural History (Rhebergen & von Hacht 2000). Rhebergen & van Kempen (2002) established the genus *Caryoconus*, by then dated as Early Silurian (Llandovery to Wenlock), as discussed below. Rhebergen (2005) also described some sponge taxa from Silurian strata exposed on Gotland, varying in age from the Lower Visby Formation (Llandovery) to the Eke Formation (Ludlow).

Although the Silurian stratigraphy of Gotland has been documented in detail in numerous publications, such as by Laufeld & Jeppsson (1976), Jaanusson *et al.* (1979), Flodén (1980), Calner *et al.* (2004), Jeppsson *et al.* (2004) and Jeppsson *et al.*

(2006), there is little information available regarding strata below the Lower Visby Formation, which is the oldest formation exposed. Sponges are either not referred to in these works or are mentioned only incidentally, as in works by Manten (1971) and Laufeld & Jeppsson (1976).

Institutional abbreviations

The following abbreviations denote the repositories of material referred to in the text:

IPPAS Institute of Paleobiology, Polish Academy of Sciences, Warszawa (Poland)

LMG Länsmuseet på Gotland (County Museum of Gotland), Visby (Sweden)

LCU Leipnitz Collection, Uelzen (Germany)

NRM Naturhistoriska Riksmuseet (Swedish Museum of Natural History, Department of Palaeozoology), Stockholm (Sweden)

PMU Museum of Evolution, University of Uppsala, Uppsala (Sweden)

SGU Geological Survey of Sweden, Uppsala (Sweden)

VCS de Vries–Collection, Sappemeer (the Netherlands)

Geological setting

The assemblage has been collected from pebble accumulations, mainly from a restricted area about 8 km in length, along the western coast of Gotland, between Gnisvärd and Stavsklint. The most productive stretch is near Nyrevsudde and Blåhäll in the Tofta parish (Fig. 2). This corresponds to the locality that yielded about 100 specimens of *Caryoconus gothlandicus* housed in the Naturhistoriska Riksmuseet (Swedish Museum of Natural History, Department of Palaeozoology), Stockholm (Sweden) and collected from the Gnisvärd/Tofta area between 1930 and 1940 by the artist Mrs. Bruce-Bendicks.

The extensive pebble accumulations in the productive area are or have been exposed to Baltic Sea breakers. They are composed of a wide range of blocks and gravel, including a range of igneous and sedimentary rock types; among them are silicified sponges, tabulate corals and stromatoporoids. Most specimens have been heavily abraded by continuous rolling (Fig. 3), and as a consequence, many surface details have been lost. Some tens of specimens were

Fig. 2. Exposed pebble accumulations at the coast near Blåhäll/Tofta. View to the North (photo: Heilwig Leipnitz, Uelzen, Germany).

Fig. 3. Detail showing the heavily worn pebbles. The arrow indicates a specimen of *Caryoconus gothlandicus* (photo: Heilwig Leipnitz, Uelzen, Germany).

collected in the Lickershamn area, and a few are from Norderstrand near Visby. Four isolated specimens are recorded from Västergarn, Mulde Fiskeläge and Kneippbyn (respectively, Heilwig Leipnitz, personal communication 2003–2010; Peter de Vries, personal communication 2011). The relatively frequent occurrence of cushion- to dome-shaped specimens of *Multistella leipnitzae* n. sp. from a restricted area near Lickershamn is remarkable, as is the absence of *Caryoconus* Rhebergen & van Kempen, 2002 at that location. This indicates at least some local variation in community ecology and localized transport from slightly different source areas.

The limited assemblage initially available was considered part of a fauna of erratic sponges of Late Ordovician age that occurs on Gotland, as well as in northern Germany and the Netherlands (Rauff 1893, 1894; von Hacht & Rhebergen 1997; Rhebergen *et al.* 2001). However, when Rhebergen & van Kempen (2002) referred one species that occurred exclusively on Gotland to the new genus *Caryoconus*, its age was found to be Early Silurian (Llandovery to Wenlock) based on acritarchs in adhering sediment. Unfortunately, one of the samples in the first series of three had been contaminated with acritarchs from a previous sample, from younger strata exposed on Gotland (Zwier Smeenk, personal communication 2007). Many additional samples have provided an accurate Late Telychian age (le Hérissé 1989), confirmed by particularly short-ranging species such as *Oppilatala insolita*, *Tylotopalla caelamenicutis* and *T. deerlijkianum*. Some of the characteristic, short-ranged species are illustrated in Figure 4.

The acritarch assemblages have been compared with those from the Når borehole 1 (le Hérissé 1989), and the Grötlingbo borehole 1 (Eriksson & Hagenfeldt 1997). In the latter paper, acritarchs from both boreholes were correlated. The acritarch taxa associated with the sponge assemblage, listed in Table 1, correspond with those occurring at depths of ca. 330–360 m. This indicates that the age of the sponges is equivalent to the *Monograptus spiralis* Graptolite Biozone (Grahn 1995). This zone corresponds to the formation that directly underlies the Lower Visby Formation. It is not exposed on Gotland, but extends below sea level along the west

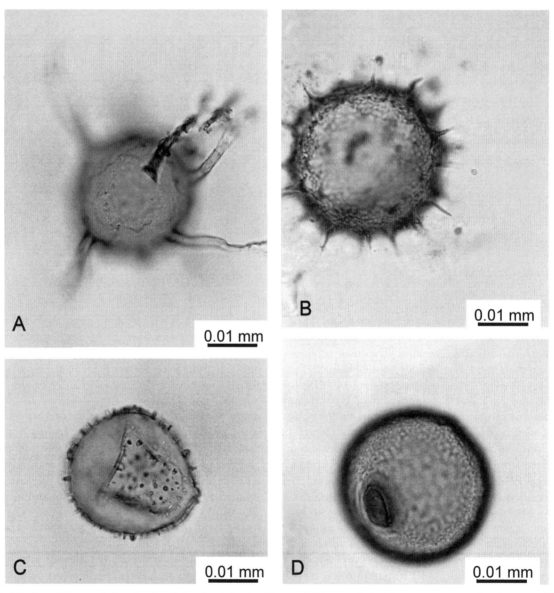

Fig. 4. Selection of characteristic acritarchs. **A**, *Oppilatala insolita* (Cramer & Diez, 1972). Scale bar 0.01 mm. **B**, *Salopidium granuliferum* (Downie, 1959). Scale bar 0.01 mm. **C**, *Visbysphaera microspinosa* Eisenack, 1954. Scale bar 0.01 mm. **D**, *Nanocyclopia* sp. Loeblich & Wicander, 1976. Scale bar 0.01 mm (photo: Dr. Zwier Smeenk, the Netherlands).

Table 1. Samples from blocks with sponges that contained acritarchs, and comparison with the depth ranges according to the Når borehole.

Sponge species	Sample	Salopidium granuliferum/S. wenlockium	Oppilatala insolita	Oppilatala grahni	Visbysphaera microspinosa	Visbysphaera brevifurcata	Visbysphaera sp. A	Nanocyclopia sp.	Membrasphaera sp.	Veryhachium cf. V. pertonensis	Tunisphaeridium parvum	Tylotopalla deerlijkianum	Tylotopalla caelamenicutis	Micrhystridium inflatum	Dictyotidium perlucidum	Dictyotidium flaviformis	Elektoriskos aurora	Buedingiisphaeridium lunatum	Ammonidium microcladum	Diexallophasis denticulata	Diexallophasis robustispinosa	Diexallophasis sp.	Cymatiosphaera aff. C. mirabilis	Dasydorus sp.
Caryoconus gothlandicus	LMG G08	X	X	X				X	X					X				X						
Caryoconus gothlandicus	LMG G16	X		X	X			X						X	X								X	
Caryoconus gothlandicus	NRM Sp10065	X			X	X		X	X															
Caryoconus gothlandicus	NRM Sp125	X						X	X									X						
Caryoconus gothlandicus	NRM Sp10066	X			X			X										X			X			
Caryoconus gothlandicus	NRM Sp10067	X						X	X					X										
Caryoconus gothlandicus	NRM Sp10069							X	X									X						
Caryoconus gothlandicus	NRM Sp10139	X			X			X	X				X	X				X						
Caryoconus gothlandicus	NRM Sp10154	X	X	X	X			X	X	X				X				X	X				X	
Archaeoscyphia minganensis	NRM Sp10074	X	X		X			X	X					X				X						
Archaeoscyphia minganensis	NRM Sp10079	X						X	X															
Archaeoscyphia gislei	NRM Sp10087	X	X					X	X					X				X			X		X	
Archaeoscyphia annulata	NRM Sp10091	X	X	X				X	X									X						
Archaeoscyphia gislei	NRM Sp10114	X						X	X									X			X		X	
Archaeoscyphia gislei	NRM Sp10078	X						X	X									X					X	
Archaeoscyphia minganensis	NRM Sp10081	X						X						X				X						
Archaeoscyphia minganensis	NRM Sp10083	X		X				X	X					X				X						
Archaeoscyphia rectilinearis	NRM Sp10132	X						X										X						
Archaeoscyphia rectilinearis	NRM Sp10117							X										X						
Archaeoscyphia gislei	NRM Sp10080	X		X				X	X	X	X	X					X	X	X	X	X		X	
Archaeoscyphia sp.	NRM 10136	X					X	X	X								X		X	X			X	
Archaeoscyphia sp.	NRM Sp10135	X	X	X	X		X	X	X					X				X		X		X		
Multistella leipnitzae	NRM Sp10094	X			X			X	X					X				X						
Multistella leipnitzae	NRM Sp10097	X			X			X	X							X		X						X
Postperissocoelia grussvardensis	NRM Sp330	X						X																
Hindia sphaeroidalis	NRM Sp10110	X	X	X				X										X						
Hindia sphaeroidalis	NRM Sp10140	X	X					X				X												
Clathrodictyon sp.	NRM Sp10129																							
Stromatoporoid	NRM Sp10137															X		X				X		
Bedrock with Pentameroides	NRM Sp10134	X			X			X	X									X						
Depth occurrence in Når-borehole (le Hérissé 1989)		326–362	322–	342–359	326–367	322–366	–	–	–	323–361	333–369	340–362	369	322–372	363–375	–361	329–	326–	327–362	324–330	328–362	–	332–362	–

coast, from Västergarn/Gnisvärd as the southernmost area, to Lickershamn in the north. In the literature, this formation is referred to as the 'Red Layer' (Lindström 1888c; Martinsson 1967). An additional indication that the Red Layer is the source of the sponge association is given by Aldridge *et al.* (1993, p. 507), stating that there is a distinct red layer immediately below the *P. celloni–P. amorphognatho-ides* Conodont biozonal boundary, which is coeval with the *Monograptus spiralis* Graptolite Biozone (Table 1). Thus, the geographical and stratigraphical position of the Red Layer and the *M. spiralis* Graptolite Biozone coincide with those areas from where the sponge association has been collected.

Although Telychian strata are not currently exposed on Gotland, occasional blocks of the host rock containing a sponge body show the sponges to have been embedded in mudstones. The host rock is identical to that preserved as matrix in superficial grooves and in spongocoels: argillaceous, yellowish-green or greyish-yellow. This is in accordance with the colouring of weathered blocks from the Red Layer (Aldridge *et al.* 1993). Some of the sponges appear as reddish bodies in a generally greyish coquina composed mainly of fragmented brachiopods.

Confirmation of the Telychian age has also been provided by the frequent co-occurrence of a single species of pentamerid brachiopod attached to tens of sponge bodies of several species. Madis Rubel (Geological Institute TUT, Tallinn, personal communication 2008) identified these as *Pentameroides subrectus* Schuchert & Cooper, 1931. This species is confined to Late Telychian formations in Estonia and Norway (Johnson *et al.* 1991) and to Middle to Late Telychian rocks on Anticosti Island, Canada (Jin & Copper 2000). Acritarchs in sample NRM Sp10134 (Table 1) from Gnisvärd, which contains *Pentameroides*, further support their coeval occurrence (Zwier Smeenk, personal communication 2001). Heilwig Leipnitz (personal communication 2008) considered the smaller specimens to be assignable tentatively to the sub-genus *Pentameroides* (*Reveroides*) sp. (Sapelnikov 1976). The latter has been reported from Ontario, northern Greenland, Urals, Altai Mountains and Baillie–Hamilton Island (Arctic Canada) and is also of Late Telychian age. This opinion is supported by other strong similarities between the sponge assemblages from Arctic Canada and Gotland discussed herein.

The very limited geographical distribution of the sponges on Gotland is also an indication of their source rock being situated nearby, that is, the Llandoverian 'Red Layer'. The fossils were probably transported by the Weichselian glaciers that shaped Gotland's present surface and adjacent seafloor.

Large blocks of strata of the Red Layer are exposed below sea level, having been glacially transported into shallower water (Lennart Jeppsson, personal communication 1999, 2004). One such was found on Skälsö near the water line, where Jeppsson sampled it. Spjeldnaes (1976) reported on an anomalous occurrence of Visby marl, 25–27 m above sea level and 4 km inland. He concluded that the huge mass of marl had been displaced by ice, probably as a frozen float.

Svantesson (1976) studied glacial striae and ice recession and described the direction of ice movement in Northern Gotland as being NNW–SSE in the northeasternmost parts and NE–SW in the northwesternmost part of Gotland. However, in a sector on the western coast, extending from Snäckgärdbaden, north of Visby, to Blåhäll/Tofta, north of Gnisvärd, the striae indicate ice movement NW–SE, perpendicular to the coast (Fig. 5). During the deglaciation, there was probably competition between two ice masses that resulted in differential ice movements in the area around and south of Visby (Svantesson 1976, figs 39, 43). Frozen blocks and slabs, originating from the Red Layer, appear to have been transported by the glacier towards the shore, or deposited on the island. Following glacial recession, these deposits were washed out and re-deposited in local accumulations on the island. When large parts of Gotland were temporarily and successively inundated during the development of the Baltic Ice Lake, Ancylus Sea and Littorina Sea, pebble accumulations were again re-deposited as gravel ridges (Fredén 1994). As a result, the sponge bodies are limited to gravel ridges eroding along the coast, which explains their precise geographical distribution. A decrease in abundance of pebbles in the Gnisvärd area and an increase in abundance in the area between Nyrevsudde and Stavsklint indicate a slow transport of the material to the north during the last fifteen years Heilwig Leipnitz, personal communication 2009).

Considering all these aspects, there is, in our opinion, sufficient circumstantial evidence to conclude that the sponge assemblage originated from the Red Layer and to establish its age as Late Telychian, stratigraphically preceding the Lower Visby Formation.

Material, repository, methods and terminology

About 150 specimens of Silurian age collected during the 19th century and the first decades of the 20th

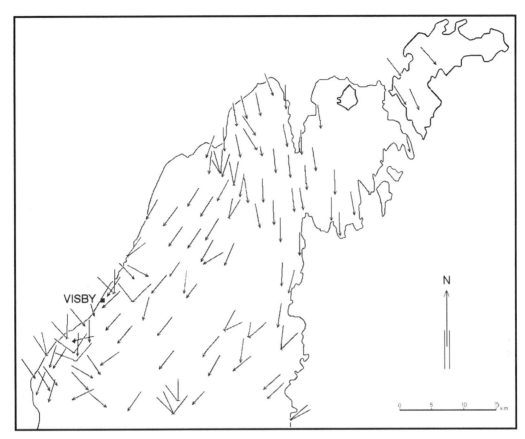

Fig. 5. Direction of ice movements in northern Gotland during the Weichselian glaciation (after Svantesson 1976, fig. 43). The limited area at the western coast, with ice direction perpendicular to the coast, coincides with the area of concentration of Telychian sponges.

century are housed in the Naturhistoriska Riksmuseet (NRM; Swedish Museum of Natural History, Department of Palaeozoology), Stockholm (Sweden), and about 40 specimens in the Länsmuseet på Gotland (LMG; County Museum of Gotland), Visby (Sweden). Minor collections are also housed in the Geological Survey of Sweden (SGU), Uppsala (Sweden), and the Museum of Evolution, University of Uppsala (PMU), Uppsala (Sweden). During the last twenty years, around 2000 specimens have been collected and were deposited primarily in the Leipnitz Collection, Uelzen (Germany), with about 1000 in the de Vries Collection, Sappemeer (the Netherlands). Type material, together with about 600 additional specimens from the Leipnitz Collection examined in this study, and eight from the de Vries Collection, has been transferred to the NRM.

In general, the skeletons and aquiferous systems in these specimens have been destroyed by intensive silicification and replacement by chalcedony. However, some specimens have been well preserved internally with transparent or translucent chalcedony surrounding the skeletal architecture. Most of the isolated sponge bodies (as well as tabulates and stromatoporoids) are coloured reddish to grey-

brown, both externally and internally, as a result of intensive oxidation. Some specimens are bluish-grey internally, with a reddish-weathered surface zone. Rare sponge bodies that are embedded in pebbles or blocks of matrix are also reddish, contrasting with the greenish bedrock, such as a specimen of *Caryoconus gothlandicus* embedded in bedrock (NRM Sp1729). Blocks of matrix incorporating sponge bodies are sedimentologically identical to matrix remnants attached to sponge bodies and consist of greenish-grey, argillaceous, silicified mudstones. The red colouration of the fossils, however, is generally limited to specimens collected in the region between Gnisvärd and Stavsklint, and those from Visby/Norderstrand, suggesting a less intense local weathering history in other areas.

Acritarchs have not been preserved in the reddish chalcedony, probably due to oxidative breakdown of carbonaceous material. The few specimens collected from the Lickershamn area are bluish-grey, and on average considerably larger than those from the Gnisvärd/Tofta area. They are similarly coloured to the more frequent Ordovician sponges in the pebble accumulations and are thus more difficult to distinguish from the latter. In addition, the composition of the assemblage is somewhat different (see below), suggesting a slightly different source area.

Many of the sponge bodies are associated with dense masses of subparallel monaxons, hexactinellid spicules and semi-articulated skeletal remains. The monaxon-rich regions may partly represent hexactinellid root tufts, but some show a clear attachment to the associated lithistid specimens, and in some cases represent a non-lithistid demosponge, *Opetionella incompta* n. sp. This type of association is uncommon among Palaeozoic sponge communities and will be discussed below.

About 60 lithistid specimens have been cut axially and/or transversely, and planes have been polished in order to study skeleton and canal architecture, where preserved. Surface details are more rarely preserved; specimens embedded in substantial blocks of matrix are exceptionally rare, and small areas of adhering bedrock are normally present only in surface furrows or cavities, such as the central spongocoel.

To establish the age of the sponge assemblages present in the Gotland pebble accumulations, parts of sponge bodies and samples of adhering matrix on 76 sponge bodies have been dissolved in hydrofluoric acid (HF). Fifty samples contained acritarchs, 30 of which were of Llandoverian (Telychian) age, and 20 of Late Ordovician age (Pirgu Stage, F_1c; Zwier Smeenk, personal communication 2001–2009); the Ordovician component is irrelevant to this study and is not considered further. In addition, samples from ten blocks of sediment that contain sponge bodies have also been dissolved in HF, but a-critarchs were absent from these due to intensive oxidation (Zwier Smeenk, personal communication 2005). The Telychian acritarch assemblage is shown in Table 1. The low numbers of acritarch specimens and taxa in samples, as well as their absence from others, may be caused by the small quantities of available adhering matrix, some samples being as small as 2 or 3 g. All samples have been deposited in the collections of the NRM, Stockholm (Sweden). Neither conodonts nor graptolites have been found in the samples.

Other fossils associated with the sponges, including tabulate corals, stromatoporoids and a single species of brachiopod, were useful in obtaining additional evidence of the Telychian age. Stromatoporoids have only been discussed if attached to a sponge body (Table 2). Harry Huisman identified the tabulate and stromatoporoid taxa (personal communication 2004–2010), some of which appear to be restricted to the Llandoverian. Stromatoporoids have been included in the present monograph, but

are given less emphasis than the demosponges and hexactinellids because identifications at species level are questionable, and distinguishing them from specimens from the Lower Visby Beds is difficult. All co-occurrences of sponges with acritarchs, brachiopods and tabulates are listed in Table 2.

The terminology of sponge spicule morphology and skeletal structures in the literature is rather confusing. De Freitas (1991) summarized numerous terms relating to Palaeozoic lithistids and added clear drawings, to propose a standardized terminology. In the present study, de Freitas' terminology has been used, supported by Figure 6, which is based on de Freitas (1991, fig. 3).

Systematic palaeontology

Phylum Porifera Grant, 1836
Class Demospongea Sollas, 1875
Order Orchocladina Rauff, 1895
Family Anthaspidellidae Miller, 1889

Genus *Archaeoscyphia* Hinde, 1889

Type species. – *Petraia minganensis* Billings, 1859.

Archaeoscyphia minganensis (Billings, 1859)

Plate 1, figures 1–8

1859 *Petraia minganensis* Billings, p. 346.

1886 *Ethmophyllum minganensis* (Billings); Walcott, p. 77, figs 6–8.

1889 *Archaeoscyphia minganensis* (Billings); Hinde, p. 143, pl. 5, figs 12–14.

1988 *Archaeoscyphia minganensis* (Billings); Rigby & Webby, p. 29, pl. 8, figs 1–6; fig. 9.

1989 *Archaeoscyphia minganensis* (Billings); de Freitas, pp. 1874–1875, figs 6M, 7B, C.

1989 *Archaeoscyphia minganensis* (Billings); Rigby & Chatterton, p. 14, pl. 2, figs 1–2; pl. 3, figs 1–2.

1995 *Archaeoscyphia minganensis* (Billings); Rigby & Desrochers, p. 17, fig. 6.1.

Material. – NRM Sp10074, NRM Sp10075, NRM Sp10079, NRM Sp10081, NRM Sp10083, NRM Sp10133, NRM Sp10142, NRM Sp10159, NRM

Table 2. Associations of two sponge species and of sponges with other fossils occurring in one block.

Porifera	Porifera																				Brachiopoda		Tabulata											
	Archaeoscyphia	Calycocoelia	Finksella.	Climacospongia	Multistella?	Postperissocoelia	Somersetella	Chiastoclonella	Hindia	Caryoconus	Lindstroemispongia	Haplistion	Warrigalia	Opetionella	Spiculite	Hexactinellida	Pachystylostroma	Clathrodictyon	Syringostromella	?Amphipora	Acritarchs	Pentameroides sp.	Alveolites sp.	Subalveolites spp.	Subalveoliella sp.	Planalveolites sp.	?Amphipora sp.	Catenipora sp.	Fossopora sp	Favosites sp.	Pachypora sp.	Heliolites sp.	Plasmopora sp	Thamnopora sp.
Archaeoscyphia							X		X									X		X	X	X	X	X			X		X			X	X	
Calycocoelia			X											X		X					X	X	X	X									X	
Finksella		X																				X												
Climacospongia					X											X						X												X
Multistella?									X	X				X							X	X				X		X				X		
Postperissocoelia																					X													
Somersetella	X								X																									
Chiastoclonella													X	X	X	X																		
Hindia	X				X		X			X								X			X	X						X		X	X	X		
Caryoconus			X		X				X												X	X	X					X		X	X	X	X	
Lindstroemispongia																							X		X									
Haplistion.													X	X										X										
Warrigalia								X				X				X								X										
Opetionella	X	X			X			X	X	X			X		X	X																		
Spiculites								X								X			X															
Hexactinellida	X	X	X	X	X			X	X	X			X	X	X						X	X	X	X				X	X				X	
Pachystylostroma																					X													
Clathrodictyon	X								X												X												X	
Syringostromella	X														X																			
?Amphipora	X	X																																

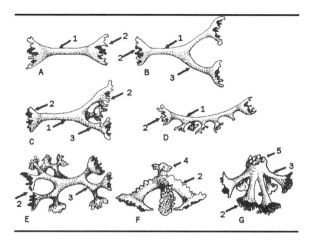

Fig. 6. Terminology of sponge spicules. **A**, amphiarborescent or I-shaped dendroclone. **B**, triclonid or Y-shaped dendroclone. **C**, polyclonid dendroclone. **D**, rhizoclone. **E**, chiastoclone. **F**, tricranoclone. **G**, spheroclone. 1, shaft; 2, zygome; 3, ray, clone or arm; 4, clad; 5, centrum. Spicules not at scale.

Sp10168 and NRM Sp10188; all collected from Blåhäll/Tofta.

Other material. – NRM Sp 10094, NRM Sp10097, NRM Sp10103, NRM Sp10200, NRM Sp10201, NRM Sp10202, NRM Sp10203 and NRM Sp10251; all from Blåhäll/Tofta.

Description. – *Archaeoscyphia minganensis* is a relatively large, obconical or cylindrical, annulated sponge. Exterior annulations vary from smoothly rounded ribs to acute horizontal crests, and the distance between them varies. In NRM Sp10075, the crests are spaced so closely that they form a collar-like outgrowth at the top (Pl. 1, fig. 1). In the present material, the largest specimen of *A. minganensis* (NRM Sp10188) is 75 mm high and 86 × 42 mm wide, compressed laterally. NRM Sp10081 is representative of the generally obconical body shape, with narrow constrictions and broad swellings (Pl. 1, fig. 2). Some adhering matrix has been dissolved in hydrofluoric acid and contained a few species of acritarchs (Table 1). In this specimen, the lateral surface shows vertically stacked lateral canals, which are separated by parieties of the skeletal mesh (Pl. 1, fig. 3). The gastral surface is visible only in a few specimens, such as NRM Sp10159 (Pl. 1, figs 4–5), and in weathered fragments, such as in NRM

Sp10168 (Pl. 1, fig. 6). Usually, spongocoels are filled with matrix or with chalcedony. A natural longitudinal section in NRM Sp10142 demonstrates that the gastral wall is roughly straight and does not follow swellings and constrictions of the sponge wall (Pl. 1, fig. 7).

The most clearly defined part of the aquiferous system in the available material is a vertically stacked radial canal array, which runs from the dermal surface, rising upwards through the sponge wall, curving downwards at about mid-wall and meeting the gastral surface at low angles (Pl. 1, figs 7–8). Canals average 0.7 × 1.0 mm in diameter and are separated vertically by 0.2–0.3 mm, by one or a few dendroclones (Pl. 1, fig. 6). Radial canals are separated horizontally by two, sometimes three parietal trabs, on average 0.30–0.45 mm wide. A system of smaller, vertical canals, roughly normal to the first set, is parallel to the skeletal trabs. They are inconspicuous in the present material, due to replacement by chalcedony.

The anthaspidellid, ladder-like trabs are pinnate, with the plane of pinnation roughly one-third of the wall thickness from the gastral surface (Pl. 1, fig. 8). The skeleton is composed predominantly of Y-shaped dendroclones in intraserial trabs about 0.4–0.7 mm long, but their shape could not generally be studied. Each is approximately 0.06–0.12 mm long.

Zygomes of dendroclones of adjacent trabs are typically supported by coring monaxons, especially in walls of intraserial canals, but have not often been preserved in the present material. However, some coring monaxons appear between zygomes of adjacent dendroclones in the gastral surface of NRM Sp10159 (Pl. 1, fig. 5).

Remarks. – It is possible that *A. minganensis* occurs frequently in this assemblage, but only a few specimens could be certainly identified due to intensive silicification and replacement of the skeleton by chalcedony, which has obscured specific characteristics.

Archaeoscyphia minganensis differs from *Archaeoscyphia rectilinearis* de Freitas, 1989, *Archaeoscyphia attenuata* de Freitas, 1989 and *Archaeoscyphia scalaria* de Freitas, 1989 by the presence of vertical canals. It differs from *Archaeoscyphia gislei* de Freitas, 1989 by its coarser spiculation, its straight, non-bifurcating radial canals, and their more regu-

Explanation of Plate 1.
Figs 1–8. ***Archaeoscyphia minganensis*** **(Billings, 1859)**
1, NRM Sp10075, showing acute ridges and crests. **2**, NRM Sp10081, with narrow constrictions and broad swellings. **3**, Lateral surface with vertically stacked radial canals separated by vertical parietal trabs. **4–5**, NRM Sp10159; **4**, oblique top view, showing part of the gastral surface; **5**, detail of Figure 4, showing trabs, with preserved coring monaxons. The arrow indicates a hexactine. **6**, NRM Sp10168, natural fracture, showing gastral surface. **7–8**, NRM Sp10142; **7**, natural median view on deep spongocoel. Gastral surfaces do not follow the swellings of the sponge wall; **8**, detail of Figure 7, showing trabs, flaring upwards to the dermal and gastral surfaces. All scale bars represent 1 cm.

lar arrangement. In *A. gislei,* radial canals are separated by one or two parietal trabs, but in *A. minganensis,* by two to three. In addition, the plane of pinnation is in *A. minganensis* near the gastral wall, but in *A. gislei* de Freitas, 1989 at mid-wall. *Archaeoscyphia annulata* (Rigby, 1973) has smooth, regular annulations and lacks the acute crests and troughs present in *A. minganensis.* In addition, *A. annulata* (Rigby, 1973) has more and longer I-shaped spicules, being more closely related to *Calycocoelia typicalis* Bassler, 1927, in which stacked radial canals are adjacent or separated by only one parietal trab. *Archaeoscyphia aulocopiformis* de Freitas, 1989 shares coarse spiculation with *A. minganensis,* but differs in its large, continuous vertical canals, which are consistently 0.4 mm in diameter, and by its widely spaced radial canals separated by usually three or four parietal trabs.

The general body shape of the Late Ordovician *A. baltica* van Kempen, 1978 is quite different from *A. minganensis,* with irregular, single or compound swellings, and reaching larger dimensions than has been reported for *A. minganensis.* The skeleton of *A. baltica* consists of longer and relatively more amphiarborescent or I-shaped dendroclones, arranged in trabs, of which the plane of pinnation is at mid-wall.

Archaeoscyphia annulata (Rigby, 1973)

Plate 2, figures 1–6

1973 *Calycocoelia annulata* n. sp. Rigby, 1973, pp. 801–804, figs 1–2.

1999 *Archaeoscyphia annulata* (Rigby); Rigby & Chatterton, p. 6, pl. 1, figs 8–10, 12–14.

Material. – NRM Sp1373, NRM Sp10091, NRM Sp10104 and NRM Sp10124; all collected from Blåhäll/Tofta.

Other material. – NRM Sp10221; from Blåhäll/Tofta.

Description. – Five tall, sub-cylindrical, weakly annulated specimens in the collections could be assigned definitely to *A. annulata.* All specimens are irregularly cylindrical, show irregular annulations and are slightly laterally compressed (Pl. 2, fig. 1). Although annulations have often been lost through abrasion, horizontal constrictions indicate the original annulations by interrupting the regularity of skeletal strands in NRM Sp1373 (Pl. 2, fig. 2). The spongocoel in NRM Sp10104 is 8 mm in diameter at the base and 13 × 20 mm at the top (Pl. 2, fig. 3). In NRM Sp10124, three annulations are developed, respectively, 23, 50 and 80 mm above the base (Pl. 2, fig. 5). Smaller annulations have probably been lost through abrasion. As a result of compression, the spongocoel at the base is only a slit and is 22 × 7 mm across at the top. Irregular growth of the sponge body is reflected in irregular patterns of the skeletal strands. Originally, the surface was covered with a cortical layer composed of numerous small dendroclones and possibly rhizoclones, which have only been preserved in narrow, sub-horizontal constrictions, as demonstrated in Figure 7 and Plate 2, figure 6. The growth pattern is somewhat aberrant, as shown in Plate 2, figure 4, where the pattern of strands is interrupted by arched surficial canals, converging to a small, lateral spongocoel, in the centre of which some vertical canals also empty, perpendicular to the surface.

The lateral exterior is marked by rows of vertically stacked, radiating excurrent canals, 0.3–0.5 mm across and separated by 0.45–0.50 mm, usually by one or two dendroclones. These are predominantly I-shaped, ca. 0.3 mm long, and the axis is 0.03–0.06 mm across. Horizontally, rows of radiating canals are separated by approximately 0.50 mm, by usually two parietal strands of smaller, mostly Y-shaped, dendroclones. Dermal spicules are preserved as casts in constriction zones and are approximately 0.06–0.10 mm long. Neither the gastral wall nor the interior pinnate skeleton could be examined due to matrix filling the spongocoel.

Remarks. – Rigby (1973) assigned the species to the genus *Calycocoelia* Bassler, 1927, but de Freitas (1989, p. 1868) referred it to *Archaeoscyphia* Hinde, 1889 on account of characteristics that were in his opinion non-specific, such as the presence of coring monaxons in attached zygomes.

Johns (1994) re-examined the type specimens of *A. annulata* and *Archaeoscyphia pulchra* (Bassler,

---

Explanation of Plate 2.
Figs 1–6. *Archaeoscyphia annulata* (Rigby, 1973)
1–2, NRM Sp1373; **1**, top view; **2**, dermal surface with slight indications of weak annulations. **3–4**, NRM Sp10104; **3**, dermal surface; **4**, detail of Figure 3, showing an irregular growth pattern that formed a new oscular centre (see text). **5**, NRM Sp10124, showing a rather irregular growth. **6**, detail of Figure 5, the cortical layer preserved in a sheltered constriction.
All scale bars represent 1 cm.

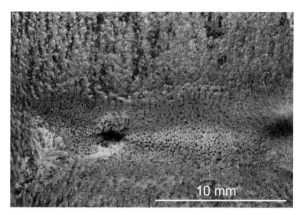

Fig. 7. Example showing preservation of dermal layer (despite abrasion) in a sheltered constriction in *Archaeosyphia annulata* (NRM Sp10124, depicted in Pl. 2, fig. 5). Note the small specimen of *Opetionella incompta* n. sp. at left.

1927) and synonymized the former with *A. pulchra*, on account of the specific differences being unobservable. His view is not followed here, due to the striking differences between the pronounced, sharp, flange-like crests of *A. pulchra* and the smooth annulations of *A. annulata*.

Calycocoelia typicalis Bassler, 1927 is distinguished from *A. annulata* by its regularly conical, goblet-shaped body form, and by having no rows, or one row, of parietal trabs between vertically stacked series of radial canals. *Archaeoscyphia alternata* de Freitas, 1989 has a regular alternation of larger and smaller, vertically stacked radial canals. *A. rectilinearis* de Freitas, 1989 has straight radial canals with a fine, reticulate skeleton. *A. scalaria* de Freitas, 1989 has much coarser spiculation and more (2–4) parietal trabs separating vertical rows of radial canals. *A. gislei* de Freitas, 1989, *A. minganensis* de Freitas, 1989 and *A. aulocopiformis* de Freitas, 1989 differ by having continuous, vertical parietal canals.

Archaeoscyphia gislei de Freitas, 1989

Plate 3, figures 1–12

1989 *Archaeoscyphia gislei* n. sp. de Freitas, pp. 1869, 1871, figs 6D, E, S, 7A.

1999 *Archaeoscyphia gislei* de Freitas, 1989; Rigby & Chatterton, p. 8, pl. 3, figs 1–7; pl. 10, fig. 9.

Emended diagnosis. – Obconical to sub-cylindrical sponge, with deep, simple spongocoel terminating at or near base of sponge, although vertical axial canals may occur in the basal part. Annulations consist of distinct, rounded, transverse ridges separated by similar rounded depressions of comparable dimensions. Radial canals consist of discrete, upward-arching structures in longitudinal and transverse section, maintaining near-constant diameter of approximately 0.45 mm throughout sponge wall. Apochetes culminate medially, and trabs, in longitudinal sections, flare upward towards both wall margins from a central surface of pinnation; skeleton typically scalariform, but uniplanar, polyclonid dendroclones account for 10 per cent of the spiculation. Vertical canals indistinct but parallel to ascending trabs. In transverse section, vertical canal openings randomly disposed, but more common in outer skeleton; radial canals slightly sinuous (emended from de Freitas 1989).

Material. – NRM Sp10072, NRM Sp 10078, NRM Sp10080, NRM Sp10087, NRM Sp10100, NRM Sp10102, NRM Sp10114, NRM Sp10120, NRM Sp10123, NRM Sp10148, NRM Sp10165, NRM Sp10166 and NRM Sp10204; all collected from Blåhäll/Tofta.

Other material. – NRM Sp10205 and NRM Sp10259; from Blåhäll/Tofta. NRM Sp10109; from Gnisvärd.

Description. – About 40 specimens could be identified as *A. gislei*, an obconical to cylindrical, annulate sponge. Annulations and depressions vary in distinctness, but usually the relation between them is rather consistent. Annulations are smooth, with near-regular separation of around 15 mm between successive crests. Depressions are approximately 2 mm deep, but annulations are less distinct due to surface abrasion.

Available specimens vary from small, slender, complete obconical specimens with a maximum

Explanation of Plate 3.
Figs 1–11. *Archaeoscyphia gislei* de Freitas, 1989
1, NRM Sp10165, lateral view. **2**, NRM Sp10123, lateral view. **3**, NRM Sp10148; oblique view on top and lateral surface. **4**, NRM Sp10120; natural median view, showing axial canals running from the base of the sponge body to the base of the deep spongocoel. **5**, NRM Sp10100; natural cross-section, showing slightly sinuous radial canals and cross-sections of parietal canals. **6**, NRM Sp10102; dermal surface with vertically stacked radial canals. **7–8**, NRM Sp10166; **7**, natural longitudinal section, showing the relation between wall thickness and spongocoel; **8**, detail of Figure 7, showing parietal canals and the plane of pinnation at mid-wall. **9–10**, NRM Sp10080; **9**, natural longitudinal section, showing radial canals; **10**, detail of Figure 9, showing the plane of pinnation at mid-wall. **11**, NRM Sp10114; well-preserved anthaspidellid structure.
All scale bars represent 1 mm.

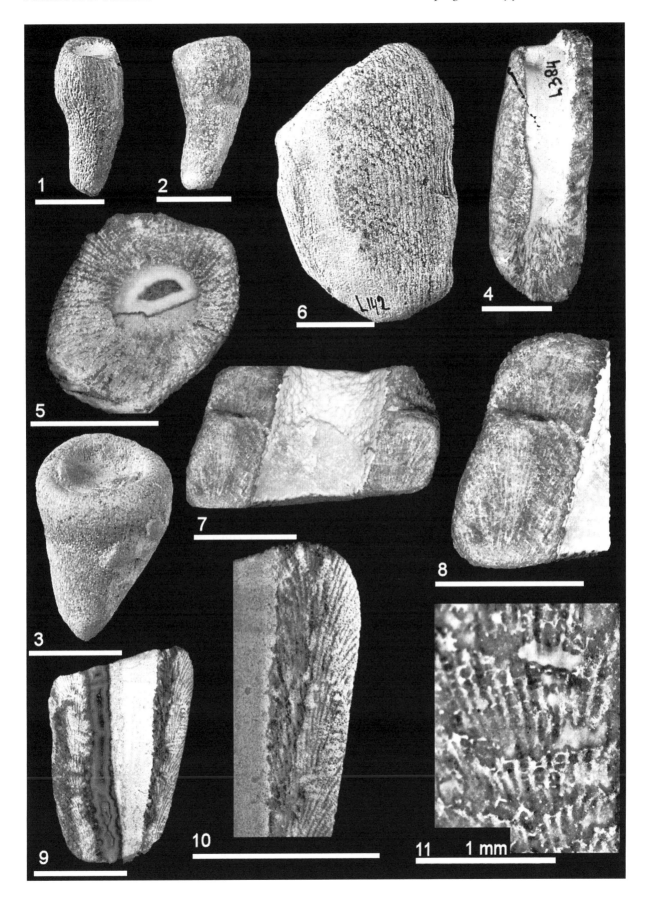

diameter of 9 mm (Pl. 3, figs 1–2), to incomplete cylindrical sponges with a diameter of at least 40 mm. Gastral surfaces are rarely visible, because spongocoels were normally filled either with chalcedony (Pl. 3, fig. 7), or with unremovable matrix (Pl. 3, fig. 9). In NRM Sp10165 and NRM Sp10123, the spongocoels extend throughout the sponge body, having a diameter of 3 mm at the base, increasing to 9 × 7 mm at the top. At half-height of the sponge bodies, the stalked lower parts show a small constriction, from where the sponge bodies expand to an obconical form. NRM Sp10148 shows a more regular, rapidly expanding form with smooth annulations. Its spongocoel is 3 mm wide at the base and 13 mm wide at the top (Pl. 3, fig. 3). Some specimens, such as NRM Sp10120 (Pl. 3, fig. 4), have a bundle of sinuous axial canals at the lower gastral surface, emptying into the base of the spongocoel.

Radial canals are 0.30–0.45 mm across, stacked in intraserial trabs and run through the sponge wall slightly sinuously (Pl. 3, figs 5, 9–10). There are six rows in 5 mm, each pair separated horizontally by usually two parietal trabs, and occasionally by one (Pl. 3, fig. 6).

Vertical parietal trabs are composed predominantly by Y- and X-shaped dendroclones, arranged pinnately, with the plane of pinnation at mid-wall (Pl. 3, figs 8, 10). They could only rarely be measured, but are 0.12 mm long and about 0.06 mm in diameter. Trabs enter the dermal and gastral surfaces at an angle of about 20°, as shown in longitudinal sections (Pl. 3, figs 7, 9).

Parietal canals occur in cross-sections as sub-circular spots of chalcedony between radial canals (Pl. 3, fig. 5). They are more abundant distally than gastrally. Large numbers of measurements were not possible, but those obtained are generally in accordance with those measured by de Freitas (1989) and Rigby & Chatterton (1999).

Clads of predominantly I-shaped dendroclones in intraserial trabs are on average 0.3 mm long and 0.06 mm in diameter (Pl. 3, fig. 11).

Remarks. – Although de Freitas (1989, p. 1869) stated that in *A. gislei*, basal, vertical cloacal apopores, common to many anthaspidellids, are absent (unlike here), NRM Sp10120 accords with other characteristics of *A. gislei*, especially the mid-wall position of the plane of pinnation. In addition, axial canals of this type have also been observed by Rigby & Chatterton (1999).

Vertical parietal canals in *A. gislei* are relatively small and somewhat irregularly arranged. Such canals in *A. aulocopiformis* de Freitas, 1989 are considerably larger and more regular and are absent from *A. rectilinearis* de Freitas, 1989, *A. scalaria* de Freitas, 1989, *A. annulata* (Rigby, 1973) and *C. typicalis* Bassler, 1927. The plane of pinnation in *A. alternata* de Freitas, 1989 is mid-wall, but the species also lacks the vertical parietal canals.

Archaeoscyphia attenuata de Freitas, 1989 has more than one cloacal opening apically, stacked radial canals are separated by two to five parietal trabs, and apochetes in one vertical row are separated by 0.5–2.0 mm. *A. gislei* has one spongocoel, its stacked radial canals are separated by one or two parietal trabs, and its apochetes in one vertical row are displaced by 0.1–1.0 mm.

Archaeoscyphia minganensis (Billings, 1859) differs from *A. gislei* by its acute and irregular annulations. It has a more regular arrangement of vertical canals, its plane of pinnation is near the gastral surface, and trabs meet the dermal surface at 50–60°. In *A. gislei*, annulations are smooth, and parietal trabs are pinnate at mid-wall and meet the dermal surface at about 20°.

Stacked radial canals in *C. typicalis* Bassler, 1927 are disposed either in adjacent rows or separated by a single parietal trab. In addition, dendroclones are generally distinctly longer than in *A. gislei*.

Archaeoscyphia rectilinearis de Freitas, 1989

Plate 4, figures 1–4

1989 *Archaeoscyphia rectilinearis* n. sp. de Freitas, pp. 1873–1874, fig. 6H, I, Q.

1999 *Archaeoscyphia rectilinearis* de Freitas, 1989; Rigby & Chatterton, p. 8, pl. 3, figs 8–11.

Explanation of Plate 4.
Figs 1–4. *Archaeoscyphia rectilinearis* de Freitas, 1989
1–2, NRM Sp10088; **1**, natural cross-section; **2**, lateral surface. **3–4**, NRM Sp10086; **3**, natural cross-section showing the regular pattern of straight lateral canals, separated by parietal trabs; **4**, lateral view of the dermal surface.
Figs 5–9. *Archaeoscyphia aulocopiformis* de Freitas, 1989
5–6, NRM Sp10144; **5**, median view of rapidly widening spongocoel with axial canals running from the sponge base to the base of the spongocoel. The left wall shows parietal trabs; **6**, oblique view on dermal surface, showing the randomly arranged apopores. 7, NRM Sp10143; gastral surface with large, vertically widely spaced apopores, separated by broad parietal trabs.
8–9, NRM Sp10164; **8**, coarse dermal surface; **9**, natural median section. The plane of pinnation is situated close to the gastral surface.
All scale bars represent 1 cm.

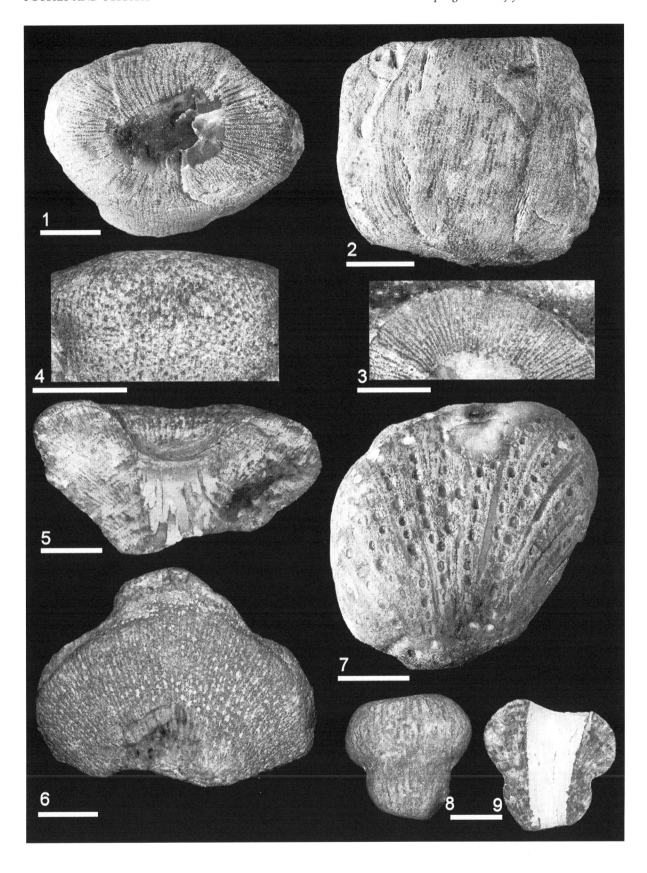

Material. – NRM Sp10086, NRM Sp10088 and NRM Sp10117; collected from Blåhäll/Tofta.

Other material. – NRM Sp10206, NRM Sp10207, NRM Sp10208, NRM Sp10209, NRM Sp10210, NRM Sp10258 and NRM Sp10262; all from Blåhäll/Tofta.

Description. – Ten specimens could be certainly identified as *A. rectilinearis*, which is a weakly annulated, obconical to cylindrical sponge with a deep spongocoel. The annulations have been lost through abrasion and are recognizable only as sub-horizontal constrictions, which retain casts of dermal tissue remnants, composed of tangential, irregularly arranged small dendroclones. All specimens are laterally compressed, and spongocoels are filled with chalcedony. In general, skeletal architecture has been preserved only in exterior regions.

NRM Sp10088 is an oval-cylindrical fragment of a large sponge, the base and top of which are missing. The spongocoel in the lower part is 23 × 16 mm and at the top 25 × 22 mm in diameter (Pl. 4, fig. 1). Numerous parietal trabs and openings of vertically stacked radial canals mark the lateral surface. Some surfaces are covered with cushion-like accumulations of monaxons. Both the cross-section and lateral surface show longitudinal fractures caused by lateral compression (Pl. 4, figs 1–2).

The anthaspidellid skeletal framework is composed of dendroclones arranged in pinnately arranged trabs. The plane of pinnation is near or at the gastral margin. Trabs flare upward and outward from the base, meeting the surface at about 30°. Vertically stacked, straight radial canals, approximately 0.30–0.45 mm across, empty into the spongocoel. At the dermal surface, they are usually separated by three parietal trabs. According to de Freitas (1989, p. 1873), canals at the gastral margin are separated by one row of parietal trabs and in mid-wall by two rows. However, gastral surfaces could not be studied in the present material, as they are obscured by chalcedony or matrix. The natural cross-section of NRM Sp10086 (Pl. 4, fig. 3) shows the reticulate skeleton and straight radial canals. The dermal surface of this specimen shows that canal openings in adjacent columns are typically alternating, rather than in horizontal rows (Pl. 4, fig. 4).

Intraserial dendroclones, usually Y-shaped, are 0.24–0.34 mm long with the axis about 0.05 mm thick. These are longer and more slender than those in parietal trabs, which are about 0.18–0.25 mm long. In tangential section, series of stacked radial canals are separated by usually two parietal trabs.

Remarks. – *Archaeoscyphia rectilinearis* is the most reticulate and regularly organized species in the genus. In *A. annulata* (Rigby, 1973), rows of radiating canals are separated by usually two parietal strands. In *C. typicalis* Bassler, 1927, straight, stacked radial canals are less regularly arranged in vertical rows and are adjacent to each other, or separated by one row of parietal trabs. In *A. scalaria* de Freitas, 1989, radial canals are sinuous and normally separated by 3–4 rows of parietal trabs; in addition, its overall spiculation is coarser. *A. alternata* de Freitas, 1989 has very distinctive large and small apochete openings alternating in each column of radial canals. Each column is consistently separated by two rows of parietal trabs. *A. minganensis* (Billings, 1859), *A. gislei* de Freitas, 1989 and *A. aulocopiformis* de Freitas, 1989 differ from *A. rectilinearis* by having vertical parietal canals, parallel to skeletal strands. *Somersetella* sp. Rigby & Dixon, 1979 differs from *A. rectilinearis* by its digitate growth form.

De Freitas (1989 described the occurrence of additional dermal skeletal elements, such as 0.11-mm-long monaxons alongside dermal dendroclones and considered them to be possibly extraneous. Citing de Freitas (1989, p. 1874): 'Dermal spicules are felted such that the specimen is rather pachydermatous, with an approximately 0.3 mm thick dermal surface'. This observation may be important in relation to the felted surfaces seen in numerous sponge bodies of nearly all taxa described in the present study.

Archaeoscyphia aulocopiformis de Freitas, 1989

Plate 4, figures 5–9

1989 *Archaeoscyphia aulocopiformis* n. sp. de Freitas, pp. 1875–1876, figs 6N, 7E.

Material. – NRM Sp10082, NRM Sp10143, NRM Sp10144 and NRM Sp10164; all collected from Blåhäll/Tofta.

Other material. – NRM Sp10211, NRM Sp10212, NRM Sp10213, NRM Sp10260 and NRM Sp10261; all from Blåhäll/Tofta.

Description. – *Archaeoscyphia aulocopiformis* possessed a relatively thick wall and a broad, deep spongocoel. Two sets of well-developed canals occur. Radial canals are sinuous, running from the dermal surface to the gastral margin. Vertical canals are large and parallel the skeletal trabs. NRM Sp10144 is a 22-mm-tall specimen, widening rapidly from 16 to

52 mm, with vertical canals measuring 1.6–2.0 mm in diameter and radial canals of 1.0–1.2 mm in diameter (Pl. 4, fig. 5). Remains of the dermal layer are preserved at the lateral surface, showing regular skeletal trabs with widely spaced apochetes (Pl. 4, fig. 6). The plane of pinnation is situated near to the gastral wall (Pl. 4, fig. 5).

NRM Sp10143 shows (Pl. 4, fig. 7) the smooth dermal margin preserved as grey chalcedony, in which numerous thin, irregularly branched dendroclones are visible, as well as some irregularly spaced apochetes of vertical canals. The gastral surface shows diverging columns of oval, vertically stacked apochetes of radial canals, 0.9–1.6 mm in diameter. Some of the apochetes appear as dark-grey spots of chalcedony, others as small hollows filled with light-grey matrix. Each column bifurcates at about half-height. There are nine columns at the base, and 17 at the top. The columns are separated by probably 2–3 vertical parietal trabs, but most spicules are not recognizable. Within a column, radial canals are spaced regularly, the vertical distance between adjacent apochete centres approximately 2.6 mm. Canals are separated vertically by 0.9 mm on average, by rows of horizontally arranged parietal dendroclones, which are poorly preserved. Large vertical canals, 0.6–0.8 mm wide, run upward and outward to the dermal surface, parallel to the parietal trabs and perpendicular to the radial canals. They are not interconnected with the radial canals (Pl. 4, fig. 7).

The skeleton is composed of small dendroclones, 0.2–0.3 mm long, with shafts that are about 0.015 thick and show elaborate zygomes at both ends. They appear as isolated spicules in translucent chalcedony, and skeletal arrangement is not preserved. Intraserial dendroclones could not be observed. NRM Sp10164, a small sponge, 28 mm high and 30 mm wide, has been cut axially. It is probably referable to *A. aulocopiformis* based on the dimensions of its wide vertical and radial canals and its coarse spiculation (Pl. 4, figs 8–9).

Remarks. – *Archaeoscyphia aulocopiformis* differs from *A. rectilinearis* de Freitas, 1989, *A. scalaria* de Freitas, 1989 and *A. alternata* de Freitas, 1989 by having vertical parietal canals. It differs from *A. minganensis* (Billings, 1859) and *A. gislei* de Freitas, 1989 in the lack of interconnection between vertical and radial canals. According to de Freitas (1989, p. 1876), *A. aulocopiformis* has a structure intermediate between the genera *Archaeoscyphia* Hinde, 1889 and *Aulocopium* Oswald, 1847. In *Aulocopium*, apochetes are at least equal in size to prosochetes, and in *A. aulocopiformis*, vertical canals are less well developed than radial ones.

Archaeoscyphia scalaria de Freitas, 1989

Plate 5, figures 1–7

1989 *Archaeoscyphia scalaria* n. sp. de Freitas, p. 1873, fig. 6A, G, R.

1999 *Archaeoscyphia scalaria* de Freitas; Rigby & Chatterton, p. 9, pl. 4, figs 1–6; pl. 6, figs 8, 9; pl. 8, figs 1–2.

Material. – NRM Sp10071, NRM Sp10099, NRM Sp10115 and NRM Sp10167; collected from Blåhäll/Tofta.

Other material. – *Archaeoscyphia* sp. cf. *A. scalaria*: NRM Sp10214, NRM Sp10215, NRM Sp10216, NRM Sp10217, NRM Sp10218, NRM Sp10219 and NRM Sp10220; all from Blåhäll/Tofta.

Description. – *Archaeoscyphia scalaria* is a low, obconical sponge, widening rapidly at an angle of about 90°, and with a deep, broad spongocoel. Apochetes in the gastral surface are arranged in a strictly scalariform manner. In the present assemblage, *A. scalaria* is a relatively rare species, represented by about ten specimens. The gastral surface of the spongocoel is visible in a natural median section of NRM Sp10099 (Pl. 5, fig. 1). The spongocoel widens regularly from the base to the top, increasing in diameter from 8 mm at the base to 15 mm at the top. NRM Sp10071 is a small, nearly complete but abraded specimen, obliquely obconical, 16 mm tall, widening from 6 mm to 28 mm (Pl. 5, fig. 2). On average, the sponge wall is 8–10 mm thick. On the dermal surface, sub-parallel parietal trabs about 0.3- to 0.4-mm-wide diverge from the base to the top, as shown in NRM sp 10115 (Pl. 5, fig. 4).

On the gastral surface, the elliptical apochetes have consistent dimensions, 0.6 mm wide and 0.9–1.0 mm long. They are arranged in strictly regular, vertical columns (Pl. 5, figs 1, 3, 5). Intraserial dendroclones are on average 0.5 mm long. Most parietal trabs are poorly preserved, except in one area of NRM Sp10167 (Pl. 5, figs 6–7). Vertical columns of apochetes commonly bifurcate in the lower half and are separated from each other by one or two parietal trabs, together on average 0.5 mm wide (Pl. 5, fig. 4). Radial canals are separated horizontally by 2–4 parietal trabs. These are sinuous and arch slightly upward from the dermal to the gastral surface, meeting the latter at an angle of about 75°. Some canals are dermally inosculate, ending blankly; others bifurcate once or twice and emerge as small

openings on the dermal surface. Dermal openings are considerably smaller, commonly 0.3 mm across (although some openings are 1 mm in diameter), and distributed rather irregularly (Pl. 5, fig. 2). The plane of pinnation is situated roughly mid-way between the gastral and dermal surfaces.

Remarks. – NRM Sp10099 is preserved in an intensely weathered state, but free of obscuring matrix (Pl. 5, figs 1–2). It demonstrates a striking similarity, both gastrally and dermally, to the specimen of Wenlockian age from the Avalanche Lake area of the Mackenzie Mountains, Canada (Rigby & Chatterton 1999, p. 34, pl. 6, fig. 9).

Archaeoscyphia scalaria is recognized by its body form, expanding rapidly from the base. It is distinct from *A. gislei* de Freitas, 1989, *A. aulocopiformis* de Freitas, 1989 and *A. minganensis* (Billings, 1859), which have horizontal, radial and vertical parietal canals. *A. rectilinearis* de Freitas, 1989 has straight apochetes and finely reticulate spiculation. *A. annulata* (Rigby, 1973) differs by its cylindrical body shape and its low number of parietal trabs, usually two, between vertical rows of apochetes. *A. alternata* de Freitas, 1989 is distinguished by its alternation of larger and smaller apochetes in one vertical row. In *C. typicalis* Bassler, 1927 vertically stacked apochetes occur in adjacent rows or are separated by one parietal trab only.

Archaeoscyphia attenuata de Freitas, 1989

Plate 5, figures 8–11

1989 *Archaeoscyphia attenuata* n. sp. de Freitas, p. 1877, figs 6B, C, P, 7F, H.

1999 *Archaeoscyphia attenuata* de Freitas, 1989; Rigby & Chatterton, p. 6, pl. 2, figs 8–12; pl. 10, fig. 2?

Material. – NRM Sp10069; collected from Gnisvärd. NRM sp10147; collected from Blåhäll/Tofta.

Description. – *Archaeoscyphia attenuata* is a stick-like, digitate sponge with multiple oscular centres. Trabs are closely spaced and are pinnate outwards, at or near the gastral surface. Two small, palmate, digitate specimens, NRM Sp10069 and NRM Sp10147, have been assigned tentatively to this species. NRM Sp10147 is severely worn and measures 33 × 20 × 14 mm. An elliptical spongocoel 5.0 × 2.5 mm in diameter is apparent at the presumed base (Pl. 5, fig. 8). From the spongocoel, skeletal trabs flare upward in all directions, at approximately 170° to the long side. A slightly compressed spongocoel measuring 3 × 5 mm in diameter approaches perpendicularly to it. The sponge body appears to have grown upwards obliquely, being wider (31 mm) than tall (20 mm), and laterally forming a spongocoel measuring 9 × 6 mm (Pl. 5, fig. 9). Numerous parietal trabs and openings of vertically stacked radial canals mark the lateral surface. The regular pattern of parallel trabs from the base upwards is modified into an array that curves strongly outwards to become nearly perpendicular to the orientation near the base. On one side, they merge into an irregular pattern of trabs, and form offshoots, with two apical spongocoels, measuring 2–3 mm in diameter. On the opposite side, radial canals are visible in cross-sections, 0.25–0.40 mm (usually 0.30 mm) in diameter and separated vertically by 0.6–1.2 mm (usually 0.9 mm). These canals are separated horizontally by 1–3 parietal trabs, which are typically 0.2 mm wide. Dendroclones in intraserial canals are usually amphiarborescent (I-shaped), on average 0.4 mm long, with axes 0.03 mm in diameter. In parietal trabs, only a few Y- and X-shaped dendroclones could be measured and are 0.25–0.30 mm long.

NRM Sp10069 is a small, irregularly digitate specimen (Pl. 5, fig. 10). The sub-circular base is 12 mm across with a spongocoel 3.5 mm in diameter. Lateral surfaces and canal systems are very irregular. The sponge body seems to be somewhat meandroid. A pattern of radially arranged trabs appears laterally at half-height between the base and visible top, probably indicating the base of a spongocoel. A second

---

Explanation of Plate 5.
Figs 1–7. *Archaeoscyphia scalaria* de Freitas, 1989
1–2, NRM Sp10099; **1**, natural median section, showing sponge wall and gastral surface; **2**, dermal surface. **3**, NRM Sp10071; oblique top view with gastral and lateral surfaces. **4**, NRM Sp10115; lateral view. **5–7**, NRM Sp10167; **5**, gastral surface with cross-section of the wall; **6**, magnified details, showing the anthaspidellid ladder-like structure; **7**, magnified detail of Figure **6**, showing the ladder-like structure of dendroclones.
Scale bars represent 1 cm on Figures **1–6**, 1 mm on Figure **7**.
Figs 8–11. *Archaeoscyphia attenuata* de Freitas, 1989
8–9, NRM Sp10147; **8**, base with small spongocoel; **9**, lateral view, showing the small spongocoel at the base and a larger one to the right. **10–11**, NRM Sp10069; **10**, lateral surface; **11**, opposite surface with two lobelike outgrowths.
Scale bars represent 1 cm.

spongocoel appears to the right (Pl. 6, fig. 11), about 6 × 3 mm in diameter, but the gastral wall is obscured by matrix that contains monaxons and hexactinellid spicules. At the top, two clusters of axial canals are visible in cross-section. One of the clusters, 4 mm across, comprises ca. 40 canals. Elliptical canals are 0.15–0.30 mm, circular canals ca. 0.20, and smaller triangular canals approximately 0.10 mm in diameter. The latter occur within parietal trabs, of which no details could be observed. Most of the dendroclones are I-shaped, 0.15–0.30 mm long, with axes about 0.045–0.060 mm thick, but Y-shaped spicules occur also.

The sponge apex is well preserved in translucent chalcedony and shows a phenomenon that is uncommon in this material: connected intraserial dendroclones, enclosing a cluster of canals in top view, are arranged such that in cross-section, they appear deceptively similar to spheroclones. This can easily lead (and, in the past, has led) to erroneous identifications. A view of the skeleton perpendicular to the upper surface shows canals and trabs with dendroclones, revealing their typical anthaspidellid structure.

Remarks. – Digitate anthaspidellids are not common (Rigby & Dixon 1979), and recognition of species is primarily based on internal architecture (de Freitas 1989). The spiculation, especially the predominance of I-shaped dendroclones, distinguishes the material here from *Somersetella* Rigby & Dixon, 1979. In *Somersetella digitata* Rigby & Dixon, 1979, skeletal trabs are more irregular and composed of Y-shaped and other polyclonid dendroclones. In addition, the digitation is more stick-like than in *A. attenuata*, based on the specimen figured by de Freitas (1989) and the palmate forms figured by Rigby & Chatterton (1999). *Somersetella amplia* Rigby & Chatterton, 1989 is considerably larger, with conspicuous lateral digitations, each with a distinct spongocoel. *Lissocoelia* Bassler, 1927 has dendroid or dichotomeous digitation and the spiculation is different. *Dunhillia* Rigby & Webby, 1988 is more slender, with numerous lateral dermal rhizoclones and relatively large gaps distributed over the lateral surface.

The irregular body shape contrasts with the regular internal spiculation. Dendroclones in *Calycocoelia* Bassler, 1927 are predominantly I-shaped, but vertically stacked radial canals are adjacent or separated by one parietal trab. In addition, radial canals in one vertical row are separated by one or two dendroclones.

Archaeoscyphia attenuata is closely related to *A. rectilinearis* de Freitas, 1989, but differs in its digitate growth form, widely spaced radial canals, less regular spicule arrangement and thinner body wall. However, these differences are subtle, and the limited material, intensive abrasion and silicification require that the assignment of the specimens to *A. attenuata* remains somewhat doubtful.

Genus *Finksella* Rigby & Dixon, 1979

Type species. – *Finksella turbinata* Rigby & Dixon, 1979.

Finksella turbinata Rigby & Dixon, 1979

Plate 6, figures 1–5

1979   *Finksella turbinata* gen. et sp. nov. Rigby & Dixon, pp. 620–622, pl. 3, figs 10–13.

2004a   *Finksella turbinata* Rigby & Dixon 1979; Finks & Rigby, p. 68, fig. 52a–d.

Material. – NRM Sp10112, NRM Sp10113 and NRM Sp10149; collected from Blåhäll/Tofta.

Description. – *Finksella turbinata* is a relatively small, turbinate to obconical sponge. The most characteristic feature is a bundle of wide axial canals, running from the base to the top. In cross-section, the canals are circular and separated from each other by skeletal parietal trabs (Pl. 6, figs 1–2). The species is represented by at least three specimens, two of which are incomplete. NRM Sp10149 is a sub-cylindrical, obconical specimen, 31 mm tall, with a maximum

width of 20–22 mm, at about two-thirds height (Pl. 6, fig. 1). At the apex, the diameter decreases to 16 mm. The apical spongocoel is 11 mm in diameter and shows a bundle of 14 circular axial canals, each 2.2–3.0 mm wide. Chalcedony has replaced nearly all structures at the base of the spongocoel. The lateral exterior part of the sponge shows long, irregularly winding and anastomosing canals, generally 0.7 mm in diameter and parallel to the surface. About 7 mm from the base, a lateral outgrowth appears, normal to the surface; it is 11 mm long and 8 mm wide, expanding slightly. If there is a spongocoel in the outgrowth, it is concealed by a coating of loose spicules. The lateral surface of the sponge body is covered with a dense cortical layer of irregular dendroclones. Casts of the dissolved spicules are 0.09–0.15 mm long. This dermal layer is preserved only in a sheltered transitional area between the sponge body and the lateral outgrowth.

The spongocoel of NRM Sp10112 (Pl. 6, figs 2–3) is 13 mm wide apically, consisting of about 12 circular axial canals, each of them 2–3 mm across. Basally, the spongocoel measures 10 × 12 mm in basal diameter, with probably nine circular canals, each 2.5–3 mm across. Axial canals in longitudinal view are shown in Plate 6, figures 4–5. NRM Sp10113 is part of a cylindrical sponge, identified by its large circular axial canals, measuring 2.8–3.0 mm in diameter. The sponge wall shows the anthaspidellid structure of upward and outward, flaring skeletal trabs. The plane of pinnation is at, or close to, the gastral wall.

Remarks. – *Archaeoscyphia* (Billings, 1859) includes several species with a bundle of axial canals, but these are restricted to the basal part and empty into the base of the spongocoel. They are distinct from *Finksella* in having more closely packed canals, which as a result are prismatic in cross-section. *Calycocoelia* Bassler, 1927 differs in that its radial canals are stacked in a strictly regular arrangement. In addition, vertical rows of radial canals are situated next to each other or separated by a single parietal row of spicules.

Species of *Somersetella* Rigby & Dixon, 1979 form lateral outgrowths or are digitate, a phenomenon that has not previously been reported in *Finksella*. In that respect, specimen NRM Sp10149 could alternatively be considered assignable to *Somersetella*. However, neither circular axial canals, rising into the top of the sponge body, nor irregularly winding, anastomosing lateral canals occur in *Somersetella*, and the specimen therefore represents the first known dendritic specimen of *Finksella*. Additional material is needed to establish whether this is a common feature

of the Gotland population, or whether the specimen represents an aberrant morphology.

Genus *Calycocoelia* Bassler, 1927

Type species. – *Calycocoelia typicalis* Bassler, 1927.

Calycocoelia typicalis Bassler, 1927

Plate 6, figures 6–10

1927 *Calycocoelia typicalis* n. gen. et sp. Bassler, 1927, pp. 390–394 (no figures).

1941 *Calycocoelia typicalis* Bassler 1927; Bassler, pp. 96–97, pl. 21, figs 3–5; pl. 24, fig. 3.

1976 *Calycocoelia typicalis* Bassler, 1927; Rigby & Chidsey, pp. 3–8, figs 1–2.

1994 *Calycocoelia typicalis* Bassler, 1927; Johns, pp. 359–365, pl. 4, figs 2–4; pl. 5, figs 2–4; pl. 6, fig. 1.

2001 *Calycocoelia typicalis* Bassler, 1927; Rhebergen, F., Eggink, Koops & Rhebergen, B., p. 80, pl. 14, figs 1–7.

Material. – NRM Sp10077, NRM Sp10101, NRM Sp10119, NRM Sp10121 and NRM Sp10132; all collected from Blåhäll/Tofta.

Description. – *Calycocoelia typicalis* is a cylindrical to slightly obconical anthaspidellid, straight or smoothly annulated in profile. Vertical rows of radial canals are arranged adjacently or separated by a single parietal trab of small dendroclones (Pl. 6, fig. 8). Five specimens among about 400 anthaspidellids studied could be certainly assigned to *C. typicalis*. The spongocoels of all specimens are filled with chalcedony, obscuring the gastral surfaces. NRM Sp10119 is the only near-complete specimen and is severely abraded (Pl. 6, fig. 6). NRM Sp10101 is the basal part of an obconical sponge body, with a spongocoel that extends from the base to the upper edge, increasing in diameter from 6 mm to 17 × 12 mm. The lateral surface shows slight annulations, crests separated by 16 mm (Pl. 6, fig. 7). On one side, there are remains of a dermal layer of obscure skeletal and aquiferous structures. The opposite surface is more intensively eroded, revealing the internal structure. The internal skeleton is composed of numerous parallel, ladder-like trabs, extending from the base and flaring upward and outward to meet the dermal surface at a high angle (NRM Sp10121, Pl. 6, fig. 10).

Vertical canals, on average 0.4 mm wide, run parallel to the trabs. They are pinnately arranged, with the plane of pinnation near the gastral surface. Radial canals, 0.5–0.6 mm in diameter, are vertically stacked and separated vertically by 1–3 dendroclones. Rows of radial canals are either adjacent horizontally or separated by one vertical parietal trab only (Pl. 6, fig. 8). These parietal trabs are smaller, generally 0.15–0.25 mm across. Canals run slightly upward from the dermal surface, reaching maximum height at mid-wall, then curve slightly downwards to meet the gastral surface (Pl. 6, fig. 10).

Dendroclones are predominantly I-shaped in *Calycocoelia* and are relatively long and slender. Clads are 0.5–0.6 mm long and 0.03–0.04 mm in diameter (NRM Sp10077; Pl. 6, fig. 9). In top view, dendroclones appear to be arranged as triangles, but this is due to seeing cross-sections of vertical skeletal trabs. Sponge bodies were originally covered with a 0.5- to 1.0-mm-thick dermal layer of small desmas, probably dendroclones and chiastoclones, forming an elaborate mesh, but this layer has been removed by abrasion.

Remarks. – Bassler (1927) gave a very brief and loose definition of the species. De Freitas (1989), in his thorough study of Silurian anthaspidellids from Arctic Canada, referred *Calycocoelia* to *Archaeoscyphia* Hinde, 1889, on account of similarities in skeletal features and aquiferous systems. Although Johns (1994) followed de Freitas' opinions in many respects, he maintained *Calycocoelia* as a distinct genus with *C. typicalis* as the type species. This opinion was based on the occurrence of exceptionally long and thin I-shaped dendroclones, not only in intraserial trabs, but also in parietal trabs. In contrast, dendroclones in species of *Archaeoscyphia* are Y- or I-shaped, and in parietal trabs usually Y- and X-shaped. In the present study, Johns' opinion is followed.

Archaeoscyphia rectilinearis de Freitas, 1989 and *A. scalaria* de Freitas, 1989 differ in the lack of vertical parietal trabs. In *A. gislei* de Freitas, 1989, the plane of pinnation is at mid-wall between gastral and dermal surfaces, and radial canals are separated by usually two parietal trabs. *A. aulocopiformis* de Freitas, 1989 has much coarser spiculation, and radial canals are separated by 2–4 parietal trabs. *A. minganensis* (Billings, 1859) is clearly annulated, with usually acute crests, and radial canals are separated by 2–3 parietal trabs. *Somersetella* Rigby & Dixon, 1979 also contains long I-shaped dendroclones, but sponge bodies are digitate. The canal system in *Finksella* Rigby & Dixon, 1979 differs from that of *Calycocoelia* in the possession of a bundle of axial canals, extending from the base to the top.

Genus *Climacospongia* Hinde, 1884

Type species. – *Climacospongia radiata* Hinde, 1884.

1884 *Climacospongia* n. gen. Hinde, p. 18.

1895 *Dendroclonella* Rauff, p. 252 (376), pl. 18, figs 3–6, 8.

1991 *Climacospongia* Hinde, 1884; de Freitas, pp. 2052, fig. 8B, D–F, G–H.

2004a *Climacospongia* Hinde, 1884; Finks & Rigby, p. 69, fig. 50, 1a–b.

Remarks. – Hinde (1884, p. 18) established *Climacospongia* to include Silurian sponges from Perry County, Tennessee, which were composed of elongate acerate spicules, radiating upwards from the base to the surface, and arranged so as to produce a series of radiating canals. He also recognized acerate spicules disposed horizontally, crossing the vertical spicules at right angles. He assigned it to the monactinellids, but Ulrich & Everett (1889, pp. 222–223) referred *Climacospongia* to the Anthaspidellidae, describing the spicules 'as four-rayed, arranged in columns, and traversed by horizontal bars', features now known respectively as zygomes of dendroclones, ladder-like trabs and shafts of dendroclones.

Rauff (1894) established *Dendroclonella rugosa*, based on specimens collected by Ulrich (1889, 1890), also from the Niagara Formation in Perry County, Tennessee, but without any reference to *Climacospongia* (see below). De Freitas (1991) synonymized *Dendroclonella* Rauff, 1894 with *Climacospongia*, although with some reservation, because he did not study the type material, but Finks & Rigby (2004a) maintained it as a separate genus. *Dendroclonella* differs from *Climacospongia* by its very fine dendroclone spiculation (on average 0.16 mm long), as well as its wrinkled dermal surface.

Incrassospongia Rigby, 1977 is distinguished by its rectangular array of diverging, vertically arranged trabs, which are cross-connected by smoothly curved, horizontally arranged trabs. According to de Freitas (1991, p. 2052), it may be synonymous with *Climacospongia*, but it lacks a clearly defined canal system parallel to the trabs, as seen in the latter. In addition, individual elements or fibres in the earliest-formed distal parts of *Incrassospongia* have a thicker diameter than equivalent dermal spicules (Rigby 1977, p. 122).

Malongullispongia Rigby & Webby, 1988 is also an acloacate sponge, but it differs in being composed essentially of Y-shaped dendroclones, having larger vertical canals, a second system of horizontal canals, and zygomes supported by coring monaxons.

Amplaspongia Rigby & Webby, 1988 has vertically radiating canals, 1.1–1.3 mm in diameter. Sub-horizontal canals are rare and discontinuous. The skeleton is composed of predominantly I-shaped dendroclones, about 0.5 mm long, of which the shafts are relatively thin, approximately 0.05–0.07 mm. Dimensions of both dendroclones and canals in *Climacospongia* are smaller.

Multistella Finks, 1960 is characterized by numerous clusters of vertical canals, which are surrounded by small, superficial, convergent canals, resulting in a stellate surface sculpture (see below). Comparable stellate structures are present on the surface of *Perissocoelia* Rigby & Webby, 1988, although the clusters of vertical canals in the latter form deep pits. Stellate structures do not occur in *Climacospongia*. The Permian *Phacellopegma* Gerth, 1927 has a surface with anastomosing grooves into which excurrent canals open.

The skeleton of *Cauliculospongia* Rigby & Chatterton, 1989 shows a similar simple structure consisting of ladder-like series of upwardly radiating trabs composed mainly of amphiarborescent dendroclones. These trabs mark a series of larger canals, 0.20 mm in diameter, and smaller ones of 0.10–0.12 mm in diameter. However, *Cauliculospongia* is small, 14 mm long and 3.2–4.0 mm wide, and twiglike, and therefore quite different to the massive, undulating skeleton of *Climacospongia*.

The canal systems in *Calycocoelia* Bassler, 1927 and *Archaeoscyphia* (Billings, 1859) differ fundamentally, with well-developed radial canals emptying into a central spongocoel.

Climacospongia undulata de Freitas, 1991

Plate 7, figures 1–6

1991 *Climacospongia undulata* n. sp. de Freitas, p. 2053, figs 6N, 8B, E.

Material. – NRM Sp10169, NRM Sp10176 and NRM Sp10189; collected from Blåhäll/Tofta.

Other material. – NRM Sp10264; from Blåhäll/Tofta.

Description. – *Climacospongia undulata* is a massive sponge without an oscular centre. The skeleton is anthaspidellid, with ladder-like trabs that radiate upward and outward, diverging at ca. 180°. Four specimens have been assigned to *C. undulata*, varying from large (NRM Sp10176, measuring 40 × 83 × 54 mm), to small (NRM Sp10189, measuring 24 × 22 × 15 mm). NRM Sp10176 has been cut vertically and polished, and its upper surface has been lightly sandpapered. This specimen demonstrates that the sponge body is composed of a cluster of cones, diverging from the base to the top (Pl. 7, fig. 2), and developing perpendicular to the distal surface. As a result, cross-sections of the cones appear as circular nodes on the upper surface (Pl. 7, fig. 1), varying in diameter from 12 to 25 mm. They originally constructed an undulose surface, but this has been abraded away. The interior angle of the cones varies between 30 and 50°. Each cone is constructed pinnately: trabs in the centre of the cone are straight, but more distally disposed trabs flare outwards to meet trabs of adjacent cones (Pl. 7, fig. 3). The increasing diameter of each cone causes distortions in the outer zones of adjacent cones; there the pattern of trabs is rather irregular, although this does not affect the diameter of the cone (Pl. 7, fig. 4).

Another conspicuous feature is repeated, irregularly undulose layering, sub-parallel to the base. The sub-parallel, concentric zones are about 0.8–2.3 mm thick and are coloured differently (Pl. 7, figs 2–3). They are separated from each other by 2 mm near the base, and up to 3.5 mm in the upper region, and maintain their separation rather consistently. Spicules could not be recognized within these zones. The layers are continuous and do not influence the development of the cones. However, they do influence zones between cones: the direction of radiating trabs below and above such a layer may change considerably, through angles of up to 50°. The undulose, layered growth has been recognized in all

---

Explanation of Plate 7.
Figs 1–6. *Climacospongia undulata* de Freitas, 1991
1–5, NRM Sp10176; **1,** view on upper surface with cross-sections of cones; **2,** median section, showing undulating, sub-parallel layered growth form and longitudinal sections of the conical structures; **3,** detail of Figure 2, showing diverging anthaspidellid trabs. Strongly undulating layers cross the cones without causing major interruptions; **4,** irregularities in the layers, in the contact plane between adjacent conical structures; **5,** detail of Figure 1, showing amphiborescent dendroclones, arranged in triangles, enclosing possibly incurrent canals. **6,** NRM Sp10169; small specimen showing conical structures and layered growth form that were developed early in sponge growth. Scale bars represent 1 cm (Figs **1–4, 6**) and 1 mm (Fig. **5**).

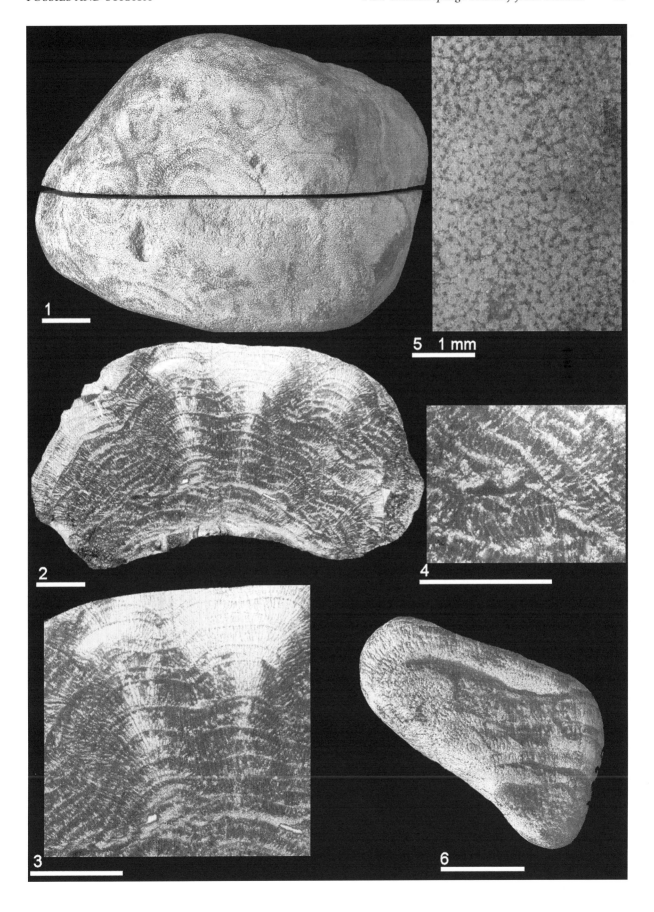

5 1 mm

specimens, including the natural longitudinal section NRM Sp10169 (Pl. 7, fig. 6).

Vertical canals are 0.30–0.35 mm wide throughout the sponge body, are always parallel to skeletal trabs and intersect the upper surface orthogonally. In top view, dendroclones are arranged in triangles, enclosing the canal outlet (Pl. 7, fig. 5). The increasing width of the sponge body, while maintaining a constant canal diameter, has been achieved by frequent bifurcation of both trabs and canals. Prosopores and inhalant canals are not clearly defined, a structure unusual among anthaspidellids, and the aquiferous system of this species is as yet poorly understood. In contrast to observations of the type material described by de Freitas (1991, p. 2053), some sub-horizontal canals, perpendicular to the vertical ones, were recognized in the distal part. They are sub-circular to elliptical and are 0.13–0.20 mm (on average 0.16 mm) across. In this area, shafts of dendroclones are recumbent.

The skeleton is composed of amphiarborescent dendroclones, usually with straight shafts, arranged in vertically stacked series and forming ladder-like trabs, although relatively few spicules have been preserved. Measurable dendroclones are 0.25–0.35 mm long, with shafts 0.04–0.05 mm thick; a few dendroclones within bifurcating trabs are smaller. Parenchymal trabs conveying additional rigidity of the skeleton are common in other anthaspidellids, but could not be observed here. Triclonid and polyclonid dendroclones seem to be absent. Monaxons, probably oxeas, occur as isolated spicules preserved within translucent chalcedony. Their position and angle of inclination with respect to neighbouring trabs indicate that they occupy their original positions, but whether they served as coring oxeas in connected zygomes could not be ascertained.

Remarks. – Two tabulates occur in the centre of the concave base in NRM Sp10176: *Catenipora* sp. occupies a surface of about 36 × 36 mm and has been overgrown in the centre by *Subalveolites* sp. or *Subalveolitella* sp. (Harry Huisman, personal communication 2009).

The occurrence of *C. undulata* in the present assemblage is the first record of the genus outside Laurentia and also the oldest representative. It is one of several species illustrating remarkable continuity of the present assemblage and the younger Silurian sponge assemblages from Arctic Canada.

Although dendroclones are generally preserved poorly, NRM Sp10176 demonstrates the skeletal structure quite clearly. The structure of individual cones, and of the intervening areas between adjacent cones, reveals additional information supplementary to that seen in the type specimens from Arctic Canada described by de Freitas (1991).

The reason for the differential colouring in the parallel banding is unresolved. The light-coloured layers consist of powdery silicified material; this occurs frequently in sponges in this assemblage and is usually related to decay of the sponge body or to later weathering. Spicules are not preserved in these light-grey bands, but these zones coincide with conspicuous changes in the direction of trabs and therefore reflect an original structural feature rather than merely chemical banding such as Liesegang rings.

The skeleton in *Climacospongia radiata* Hinde, 1884 is coarser, and dendroclones are on average larger (0.65–0.68 mm; de Freitas 1991) and are interconnected by coring oxeas. It lacks the conical nodes seen in *C. undulata*.

Climacospongia snowblindella de Freitas, 1991 is 'a pachydermatous, cloacate, thin-walled sponge with a deep, broad spongocoel. (…) In tangential section, polygonal, elliptical, or sub-quadrate skeletal gaps are not arranged in vertical series' (de Freitas 1991, p. 2053), therefore differing in several critical respects from *C. undulata*.

Genus *Multistella* Finks, 1960

Type species. – *Multistella porosa* Finks, 1960.

Remarks. – Finks (1960) established *Multistella* for Permian sponges from the Guadalupian Limestone Member of the Cherry Canyon Formation in the Guadaloupe Mountains (Texas). To our knowledge, it has not previously been reported from elsewhere.

Amplaspongia Rigby & Webby, 1988, from the Upper Ordovician Malongulli Formation (New South Wales, Australia), includes massive, hemispherical to globular sponges with a canal system different from that of *Multistella*. Major canals, paralleling the skeletal trabs, can be isolated or clustered, with the latter arrangement resulting in stellate structures; (sub-) horizontal canals are rare or absent, in contrast to the numerous, sub-horizontal small canals that empty into the oscular clusters of *Multistella*.

Climacospongia Hinde, 1884 is a massive sponge that differs from *Multistella* by its predominantly I-shaped dendroclones, and its skeletal arrangement of anthaspidellid trabs with large, deep cones; this differs from the stellate surface structures of oscular centres in *Multistella*.

Perissocoelia Rigby & Webby, 1988, from the Upper Ordovician Malongulli Formation (New South Wales, Australia), differs from *Multistella* in having numerous deep oscular pits distributed over the sides and upper surface. In addition, the strand-forming arrangement of dendroclones in *Perissocoelia* is much more irregular than it is in *Multistella*.

Hydraspongia Rhebergen, 2007 is a Late Ordovician sponge occurring in Pliocene, and Early to Late Pleistocene, fluvial deposits in Germany and the Netherlands, respectively. On Gotland, it occurs in Late Pleistocene deposits. They all originate from a north-eastern Baltic source area. *Hydraspongia* differs by its conspicuous growth increments and its bulbous oscular heads, which were formed continuously; each of these bulbous structures shows converging canals and a deep oscular pit.

Postperissocoelia n. gen. differs in its possession of oscular heads (expanded apical regions housing the oscula) and in its skeleton having didymoclone-like dendroclones (see below).

Phacellopegma Gerth, 1927, a Permian compound sponge from Timor re-described by Finks (1960), is a massive, sub-hemispherical sponge with a surface covered by deep anastomosing grooves, into which circular excurrent canals open.

Multispongia Carrera, 2006 is a compound, lamellar sponge from Lower Ordovician deposits in the Argentine Precordillera. Coalescent interconnected spongocoels form a thick, bulbous, laminated body. Associated spongocoels are deep, reaching almost to the base of the sponge. The skeletal structure is composed mainly of trabs, in which vertical, fused coring monaxons are the main elements. These trabs are locally interconnected by I-shaped dendroclones. Thus, the genus differs markedly from *Multistella*.

Anthaspidella Ulrich & Everett *in* Miller, 1889 includes several species, all of which typically have a concave, saucer-like top, in which a number of oscular centres open. All species have a stalked base, but lack both a dense basal skeletal layer and alternating zones within the dermal layers.

Irregularly hemispherical specimens of *Chiastoclonella* sp. may show superficial similarities to *Multistella*, including overhanging layers, but they are composed predominantly of chiastoclones that are very irregularly shaped and are distinctly smaller than dendroclones (see below). As a consequence, neither a regular arrangement of skeletal trabs nor a well-developed canal system are present.

Multistella leipnitzae n. sp.

Plate 8, figures 1–7

Derivation of name. – Named in honour of Mrs. Heilwig Leipnitz (Uelzen, Germany), who collected nearly all of the material discussed herein, including most of the specimens of this new taxon.

Holotype. – NRM Sp10107 (Pl. 8, figs 1–3); transported erratic specimen from gravel accumulations at the coast at Blåhäll/Tofta (Gotland, Sweden); coordinates 57°31′ N; 18°06′ E.

Paratypes. – NRM Sp10093; transported erratic specimens from the coast near Stenkyrkehuk (Gotland, Sweden) and NRM Sp10179; transported erratic specimen from Tofta (Gotland, Sweden).

Other material. – NRM Sp10127 and NRM Sp10191; collected from Gnisvärd. NRM Sp10128 and NRM Sp10224; from Stenkyrkehuk. NRM Sp10191, NRM Sp10192, NRM Sp10193, NRM Sp10194, NRM Sp10195, NRM Sp10196, NRM Sp10222, NRM Sp10223 and NRM Sp10268; from Blåhäll/Tofta.

An additional 20 unnumbered specimens are in the Leipnitz Collection, 14 of which are from Blåhäll/Tofta, three from Gnisvärd, one from Lickershamn and two from Stenkyrkehuk, as well as one unnumbered specimen from Gnisvärd in the de Vries Collection.

Repository. – The type- and other numbered material are housed in the Swedish Museum of National History (NRM), Department of Palaeozoology, Stockholm, Sweden.

Diagnosis. – Massive sponge with concentrically wrinkled dermal layer on a usually flat base. Stellate oscular centres distributed over smooth, convex surface. Clusters of shallow, slightly winding and distally branching canals run parallel to surface, converging normal to surface in a cluster of apopores of excurrent canals. Isolated, sub-circular apopores positioned between stellate centres. Relatively open skeletal structure of amphiarborescent dendroclones arranged in regular series of stacked trabs.

Description. – *Multistella leipnitzae* n. sp. is a massive, sub-hemispherical to ovoid sponge, with considerable variation in size and growth form. The dimensions of the holotype, NRM Sp10107, are 67 × 45 × 31 mm (Pl. 8, fig. 1), but the largest specimen known, NRM Sp10093 is 105 mm long,

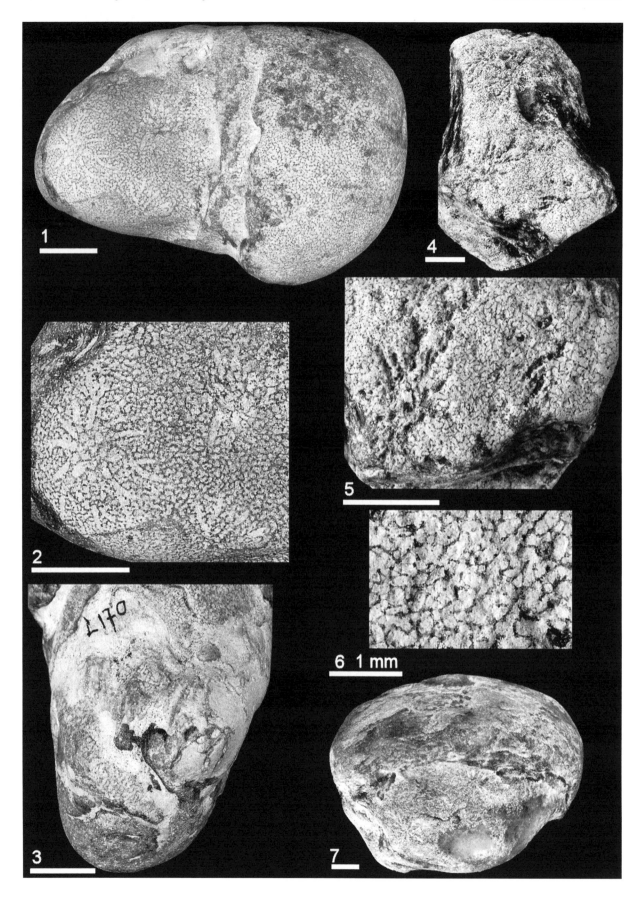

84 mm wide and 55 mm high (Pl. 8, fig. 7). The smallest complete specimen known, NRM Sp10195 is 33 × 22 × 19 mm. The base is generally flattened, as in NRM Sp10093, but some specimens, such as the holotype, are obconical and stalked (Pl. 8, fig. 3). Usually, the base shows a concentrically wrinkled layer.

The smooth, convex upper part is formed by layers that overhang earlier-formed layers, sometimes including the base, and is comparable to a stromatoporoid growth form (e.g. NRM Sp10093). Stellate oscular centres are distributed over the entire upper surface and convex sides (Pl. 8, figs 1–2, 4–5). Each of the centres consists of a cluster of three to eight vertical canals, which converge slightly from the lower part of the interior, approaching the surface normally. Each cluster is surrounded by four to ten radiating canals, parallel to the surface. These canals are usually 1–2 mm wide and 4–9 mm long, slightly sinuous, occasionally branching distally (Pl. 8, fig. 2). The diameter of a stellate oscular centre averages 16 mm. The distance between clusters varies from 9 to 20 mm, but is usually ca. 12 mm. No examples of interconnection of radiating canals from adjacent centres were seen. Some clusters form shallow depressions, ca. 1–2 mm deep and ca. 3 mm across, but the stellate centres can usually be recognized only by differences in colouring. In heavily abraded specimens, the superficial canals have been lost, but clusters of vertical canals demonstrate their original positions. Endosomal canals, ca. 0.4–0.6 mm wide, run parallel to the skeletal series from the basal region towards the surface, emptying normal to the surface, in areas between the stellate structures. Entire examples of these could only be traced occasionally, due to the poor preservation of the interior. Smaller canals are also present, 0.2–0.3 mm wide, running normal to the skeletal strands, but are only occasionally visible.

The anthaspidellid skeleton is composed predominantly of stacked, relatively slender, amphiarborescent dendroclones, which are arranged in trabs that radiate from the skeletal centre at the base. The dendroclones are 0.2–0.5 mm, usually 0.4 mm long, with shafts ca. 0.02–0.03 mm across. Their elaborate zygomes are relatively loosely constructed. Interconnected zygomes appear as small dots (Pl. 8, fig. 5–6). Trabs meet the surface approximately at right angles, creating a characteristic surface pattern of interconnected shafts arranged parallel to the surface, forming a mesh of small triangles enclosing skeletal pores, which are on average 0.45 mm wide (Pl. 8, fig. 6). On the polished surfaces of NRM Sp10194, some isolated pyritized dendroclones in translucent chalcedony are measurable, together with some connected dendroclones in distal areas of NRM Sp10196.

Remarks. – *Multistella leipnitzae* n. sp. is rather similar to the Permian *M. porosa* Finks, 1960, sharing the massive, sub-spherical body form and similar primary skeletal structure. However, the dendroclones in *M. leipnitzae* are more slender, their shafts being longer and thinner. In addition, their zygomes are less complex and more open, compared with the stockier dendroclones with knotlike zygomes figured by Finks (1960) and Mehl-Janussen (1999). The general form of the dendroclone in *M. leipnitzae* accords with those of other Early Palaeozoic Anthaspidellidae and differs from those of Middle Palaeozoic Anthaspidellidae. Erection of a new genus may be appropriate in the future, given better preserved material that clarifies the potential differences.

Co-occurrences with other taxa appear relatively more frequently than on other anthaspidellids. On the convex upper surface of NRM Sp10093, *P. subrectus* has settled and has been partly overgrown by the sponge. NRM Sp10107 nucleated on *Catenipora llandoverensis* or on an adjacent specimen of *Pentameroides*. NRM Sp10191 is part of a dense association comprising *M. leipnitzae* n. sp., *Caryoconus gothlandicus* (Schlüter, 1884), *P. subrectus* and *Plasmopora* cf. *P. rudis*. Cushion-like accumulations of monaxons, such as on NRM Sp10193, as well as hexactinellid sponge spicules, occur on the surface or in furrows of several specimens.

Explanation of Plate 8.
Figs 1–7. *Multistella leipnitzae* n. sp.
1–3, NRM Sp10107; holotype. Erratic from gravel accumulations at the coast at Blåhäll/Tofta (Gotland, Sweden); **1**, view of upper and lateral surface. Stellate structures are distributed all over the surface; **2**, detail of Figure 1, showing three clusters of axial apopores surrounded by converging, stellately arranged surficial canals; **3**, base of the holotype with *Pentameroides subrectus*, overgrown by the sponge. **4–6**, NRM Sp10179; paratype; **4**, view of dermal surface with weathered surficial canals; **5**, detail of Figure 4, with cluster of axial canals and converging surficial canals; **6**, detail of Figure 5, showing amphiarborescent dendroclones. **7**, NRM Sp10093; paratype. Erratic from pebble accumulations at the coast near Stenkyrkehuk (Gotland); view of upper surface. The layered construction is demonstrated in the lower part, where the outer layer has been lost, exposing an underlying layer with several stellate structures. The dark spot at the top is a chalcedonized specimen of *Pentameroides subrectus*.
Scale bars represent 1 cm (Figs **1–5, 7**) and 1 mm (Fig. **6**).

Genus *Somersetella* Rigby & Dixon, 1979

Type species. – *Somersetella conicula* Rigby & Dixon, 1979.

1979   *Somersetella* n. gen. Rigby & Dixon, p. 614.

1989   *Archaeoscyphia* Hinde, 1889; de Freitas, pp. 1861–1868.

1989   *Somersetella* Rigby & Dixon, 1979; Rigby & Chatterton, pp. 31–32.

1999   *Somersetella* Rigby & Dixon, 1979; Rigby & Chatterton, p. 10.

2004a   *Somersetella* Rigby & Dixon, 1979; Finks & Rigby, p. 96.

Remarks. – The genus *Somersetella* was erected by Rigby & Dixon (1979) to include digitate anthaspidellid sponges with dermal specialization, other than a basal imperforate layer. In addition, they considered the occurrence of oxeas in canals (not coring spicules in connected zygomes within series of dendroclones) to be a diagnostic feature of *Somersetella* and *Climacospongia* Hinde, 1884. Rigby & Chatterton (1989, p. 32) observed that dermalia and coring monaxons were not as distinct as described by Rigby & Dixon (1979). De Freitas (1989) confirmed this opinion, although he did not consider the digitate growth form to be taxonomically critical and synonymized *Somersetella* with *Archaeoscyphia* (de Freitas 1989, p. 1868). To Rigby & Chatterton (1999, p. 10), it seemed, 'advantageous, however, to differentiate digitate from related non-digitate sponges taxonomically for the present'. This may explain the contradiction that the non-digitate *Somersetella conicula* still exists as the type species of a genus that comprises mostly digitate sponges. In the present study, the opinion of Rigby & Chatterton (1999) has been followed.

Somersetella amplia Rigby & Chatterton, 1989

Plate 9, figures 1–3

1989   *Somersetella amplia* n. sp. Rigby & Chatterton, pp. 33–34, pl. 6, figs 8–11.

Material. – NRM Sp10155; from Blåhäll/Tofta.

Other material. – NRM Sp10225, NRM Sp10226, NRM Sp10227 and NRM Sp10228; all collected from Blåhäll/Tofta.

Description. – *Somersetella amplia* is a digitate sponge with a spongocoel of diameter roughly one-third of the main sponge body or branch. This is demonstrated in NRM Sp10155, which is the lower part of a digitate sponge, 43 mm tall, and 25 mm wide (Pl. 9, figs 1–3). The obconical main body has a central axial spongocoel, which is circular at the base, 5 mm in diameter, and elliptical at the broken top, measuring 11–15 mm in diameter (Pl. 9, fig. 1). Two branches developed simultaneously, next to each other and laterally fused. The larger branch is 20 mm in diameter and its spongocoel is 6 mm across. The orientation of the secondary spongocoel is 120° to the main spongocoel. The smaller branch is ca. 18 mm across, with a spongocoel 4–5 mm in diameter, and at an angle of about 50° to the main spongocoel (see arrow in Pl. 9, fig. 3).

Gastral surfaces could not be examined, as the spongocoels are filled with partly agatized, concentrically layered chalcedony. Remains of a thin dermal layer appear in narrow, sub-horizontal constrictions, which are in turn largely covered by an extraneous layer of monaxial spicules.

The dermal surface (Pl. 9, fig. 3) shows a consistent pattern of vertically stacked radial canals, about 1 mm apart. They are approximately 0.3 mm in diameter and separated within one row by domi-

---

Explanation of Plate 9.
Figs 1–3. *Somersetella amplia* Rigby & Chatterton, 1989
1–3, NRM Sp10155; **1**, view of natural cross-section of the main sponge body; **2**, oblique view of natural cross-sections of two offshoots. The base of the main sponge body is at lower left; **3**, lateral view of outer surface of the main sponge body, with strictly regular arrangement of prosopores. The arrow indicates the position of the right spongocoel shown in Figure 2. The grey patch next to the arrow is a specimen of *Opetionella incompta* n. sp.
Figs 4–8. *Somersetella digitata* Rigby & Dixon, 1979
4–5, NRM Sp10122; **4**, lateral view, showing two spongocoels; **5**, the opposite of Figure 4, showing the conical branch with dermal surface. 6–8, NRM Sp10186; **6**, lateral view of digitate sponge body; **7**, one of the branches, showing the dermal surface; **8**, natural cross-section of one of the spongocoels with sinuous radial canals separated by parietal trabs. The spongcoel is composed of a bundle of axial canals.
All scale bars represent 1 cm.

nantly amphiarborescent, horizontally arranged dendroclones. Rows of canals are usually separated by three parietal trabs. Horizontally, ten rows of canals occur in 10 mm. In cross-section, radial canals appear as white, slightly sinuous lines representing infills of white chalcedony (Pl. 9, fig. 1). They are separated by parietal trabs, many of which bifurcate to form an intercalated trab. Canals are usually separated gastrally by two parietal trabs and by 3–4 trabs dermally. Vertical canals parallel the upward-and-outward divergent skeletal trabs and occur in two sizes: larger canals are 0.15 mm in diameter and smaller ones 0.06 mm.

Intraserial dendroclones in radial canals are predominantly I-shaped, on average 0.15 mm long, and with shafts 0.06 mm across. Dendroclones in parietal trabs are commonly Y- and X-shaped and on average 0.10 mm long. Some small rhizoclones are visible in the walls of vertical canals. The dermal layer is preserved as moulds of tangentially arranged, irregularly connected dendroclones, approximately 0.10 mm long.

Remarks. – Part of the surface of NRM Sp 10155 is overgrown by *O. incompta* n. sp., as shown in Plate 9, figure 3, next to the arrow.

Although *S. amplia* is represented by only a few, incomplete specimens, it is readily distinguishable from *S. digitata* Rigby & Dixon, 1979 by its larger dimensions, coarser texture, the size and spacing of the canals and its thinner dermal layer (Rigby & Chatterton 1989, p. 34). In addition, dendroclones in the dermal layer of *S. amplia* are on average smaller than those in *S. digitata*. *Somersetella conicula* Rigby & Dixon, 1979 differs in being non-digitate (see above).

Somersetella digitata Rigby & Dixon, 1979

Plate 9, figures 4–8

1979 *Somersetella digitata* n. sp. Rigby & Dixon,
     p. 618, pl. 2, figs 1–2.

1989 *Somersetella digitata* Rigby & Dixon, 1979;
     Rigby & Chatterton, p. 32, pl. 7, figs 1–4.

1999 (?)*Somersetella digitata* Rigby & Dixon,
     1979; Rigby & Chatterton, p. 10, pl. 4, fig. 7.

Material. – NRM Sp10122 and NRM Sp10186; collected from Blåhäll/Tofta.

Other material. – NRM Sp10229.

Description. – *Somersetella digitata* is a digitate sponge with sub-cylindrical branches. Each branch has a deep spongocoel throughout its length. *S. digitata* is represented by four, possibly six, incomplete, abraded specimens. Annulations have not been observed. Due to the erosion to which they have been exposed, the thinnest branches have been lost and only their bases are preserved. Nevertheless, most of the specimens show characters that make identification reliable.

NRM Sp10122 has two branches. One is 13 mm in diameter and appears in cross-section near its base (Pl. 9, fig. 4). It has a 3-mm-wide central spongocoel, in which four or five axial canals, 0.3–0.5 mm in diameter, can be seen. The other offshoot appears as a regular cone, 21 mm tall and 13 mm wide at the top, with a spongocoel measuring about 4 mm in diameter (Pl. 9, fig. 5).

NRM Sp10186 has an unusual body form with a more-or-less flattened, slightly convex base and three offshoots. One is vertically oriented, and two others are disposed sub-horizontally (Pl. 9, fig. 6). The vertical offshoot is sub-cylindrical, 14 mm in diameter and widens near the base. It has a 6 × 8-mm-wide spongocoel. The two perpendicular offshoots each form a cylinder, about 13 mm in diameter with a sub-circular spongocoel, 7 mm in diameter (Pl. 9, fig. 6). Each spongocoel consists of a bundle of axial canals, which are 0.8 mm in diameter. Radial canals, 0.45–0.50 mm by 0.30 mm in diameter, are regularly stacked in intraserial trabs (Pl. 9, figs 5, 7); they are slightly sinuous and run gently upwards from the dermal to the gastral surface. They are separated horizontally by one, sometimes two, 0.20- to 0.25-mm-wide parietal trabs (Pl. 9, figs 4, 8).

The dermal layer has been uncovered by fine air-pressure preparation and shows (Pl. 9, figs 6–7) a coarse mesh of diagenetically modified dendroclones. Small, sinuous, sometimes digitating canals, 0.5 mm wide and 2–4 mm long, are visible along the lateral surface, partly within and partly on the surface of the dermal layer. They are oriented sub-parallel to the skeletal strands, but some run at an angle of about 45° to the vertical trabs; it is difficult to recognize where they start and end. They are possibly remains of tiny canals within the cortical layer. Individual dendroclones could not been measured.

Remarks. – The top of one of the branches of NRM Sp 10186 is overgrown by a tabulate coral, probably *Alveolitella* sp.

Somersetella digitata differs from *S. amplia* Rigby & Dixon, 1979 by its generally smaller dimensions, its finer texture and conspicuous regularity in canal systems and skeletal structure, the size and spacing of the canals, and its thick dermal layer. In addition, dendroclones in the dermal layer of *S. digitata* are on average smaller than those in *S. amplia*. *Somersetella conicula* Rigby & Dixon, 1979 is a non-digitate sponge and is thus easily distinguished from *S. digitata* (explained above).

De Freitas (1991, fig. 8G) described and figured a poorly preserved, digitate, probably fan-like anthaspidellid sponge from the Cape Phillips Formation (Ludlow) on Cornwallis Island (Arctic Canada). It is more-or-less flat-based, with eleven (five distinct) branches, each with a spongocoel. He assigned it tentatively to *Climacospongia* Hinde, 1884, although sponges in this genus are acloacate and not digitate. The specimen is quite similar in shape and dimensions to NRM Sp10186, which also shows a more-or-less flat base from which branches develop. The Canadian specimen is probably better assigned to *S. digitata* than to *Climacospongia*.

Family Streptosolenidae Johns, 1994

Genus *Postperissocoelia* n. gen.

Type species. – *Perissocoelia? spinosa* Rigby & Chatterton, 1989.

Derivation of name. – Contraction of *post* (Latin) meaning later in time, and the closely related Ordovician genus *Perissocoelia*.

Included species. – *Postperissocoelia spinosa* (Rigby & Chatterton, 1989) and *P. gnisvardensis* n. sp.

Diagnosis. – Stalked to massive anthaspidellid sponges with dense dermal layer around the base. Numerous conical oscular depressions, either on convex crest or at centres of oscular heads; associated radiating canals mark upper surface and occur in interior as stacked, arcuate, radiating canals, sloping steeply downward and outward in the lower part of sponge. Base of each oscular pit with cluster of excurrent canals. Discontinuous trabs radiate upward from skeletal centre near base. Principal spicules arranged in ladder-like series of dendroclones. Skeleton distally composed of didymoclone-like dendroclones, zygomes of which are fused, forming sub-globular knots, some of which double-connected by sub-parallel clones. Knots may be spinose due to sharp oxeas placed normal to the surface. Small rhizoclones may occur irregularly throughout the sponge.

Remarks. – The new genus *Postperissocoelia* has been proposed herein for compound anthaspidel-lid sponges in which the skeleton is composed distally of didymoclone-like dendroclones. This feature is not known among Ordovician anthaspidellids.

Perissocoelia was established by Rigby & Webby (1988) for Late Ordovician sponges from the Malongulli Formation, New South Wales (Australia), which are characterized by numerous osculi distributed over the upper and lateral surfaces. The skeleton is composed predominantly of common dendroclones, which also differs from *Postperissocoelia*.

Hydraspongia Rhebergen, 2007 includes compound sponges that usually form bulbous oscular heads and show conspicuously distinct growth stages. *Postperissocoelia* n. gen. resembles *Hydraspongia* in general form, but the latter is composed of dendroclones and chiastoclones, and didymoclone-like spicules are absent.

Phacellopegma Gerth, 1927, re-described by Finks (1960), is a sub-hemispherical compound sponge from the Permian of Timor. It is a massive sponge that differs from *Postperissocoelia* in having a surface covered by deep, anastomosing surface grooves, into which circular excurrent canals open.

Multistella Finks, 1960, from the Permian in Texas, is a sub-hemispherical sponge with a layered growth form, comparable to that of stromatoporoids, in which new layers overlap previously formed ones. The surface bears small, astrorhizae-like oscular centres, only 1–2 mm deep, into which empty small surficial canals. *Multistella* does not develop oscular heads.

Anthaspidella Ulrich & Everett *in* Miller, 1889 comprises several species, all of which differ from *Postperissocoelia* in possessing a stalked basal part and a concave, saucer-like top, on which there are several cloacal oscules.

Perissocoelia? gelasinina Rigby & Chatterton, 1989, also a Silurian sponge from Arctic Canada, differs from *Postperissocoelia* n. gen. in lacking distinct growth stages and bulbous heads. In addition, its canals are extremely short and very irregular, and didymoclone-like dendroclones do not occur.

Modification of dendroclones into didymoclone-like forms appears to be have developed in the Anthaspidellidae only in the Early Silurian. *Postperissocoelia* n. gen. is proposed to include taxa that demonstrate this development and to exclude those with simple dendroclones, such as the Ordovician *Perissocoelia* Rigby & Webby (1988). As a consequence, *Perissocoelia? spinosa* Rigby & Chatterton, 1989 is here reassigned to the new genus *Postperissocoelia*.

Postperissocoelia gnisvardensis n. sp.

Plate 10, figures 1–3

Derivation of name. – Named after the village of Gnisvärd (Gotland, Sweden; Fig. 1), from where this specimen and numerous other Telychian erratic sponges have been collected.

Holotype. – NRM Sp330, holotype (Pl. 10, figs 1–3); from an erratic block collected near Gnisvärd, Gotland.

Material. – One specimen; the specimen has been cut longitudinally. A small area on one of the sides has been cut in order to be dissolved in hydrofluoric acid.

Repository. – NRM Sp330 is housed in the Swedish Museum of Natural History, Department of Palaeozoology, Stockholm.

Diagnosis. – Compound sponges with two or more growth stages with apical oscular heads. Each head with a small, shallow spongocoel, into which short, converging excurrent canals empty. In sub-dermal parts, zygomes of adjacent dendroclones fused to spherical knots, which frequently are double-connected by didymoclone-like dendroclones.

Description. – The nearly complete, massive, elliptical sponge is 96 mm long, 45 mm wide and 63 mm high. The upper part is damaged, with remains of lateral heads only. Longitudinally, the sponge body is composed of three distinct parts. The obconical basal part is 20–25 mm high and covered by a wrinkled, slightly annulated, dermal layer, which is partly eroded (Pl. 10, fig. 2). The sides meet the base at an angle of approximately 150 and 40°, respectively. A wreath-like exposure of the dermal layer marks the transition between the basal and middle parts. The latter is composed of a series of six or seven elliptical oscular heads, each of which is approximately 32–34 mm long and 26–30 mm wide (Pl. 10, fig. 2). Their original heights are unknown, because the apices are absent. More heads were presumably developed, but they have been overgrown in the next stage of growth (see below).

Another wreath-like development of the dermal layer marks the transition from the middle to the upper part, which is composed of a row of nine oscular heads. They are arranged rather regularly in a more-or-less horizontal plane, forming the crest of the sponge (Pl. 10, fig. 1). These heads are on average smaller than those in the middle part, 20–24 mm in diameter; however, their original heights are also unknown, because their apices are broken. Nearly all of them show remains of the internal canal system (Pl. 10, fig. 3). Each of the oscular heads has its own system of excurrent canals. In the centre of the best-preserved head, cross-sections of a cluster of six vertical, axial canals, as well as oblique cross-sections of radially arranged canals, can be observed (Pl. 10, fig. 3, arrows). The radiating canals are short, measuring 0.3–0.5 mm wide and 3–4 mm long.

Numerous tiny holes over the entire surface mark weathered and dissolved skeletal elements. The anthaspidellid skeleton can be studied in small areas only, most of them at or just below the dermal surface. The framework is composed of skeletal strands radiating from the base to meet the surface at about 80–90°. Dendroclones are preserved only as ghosts in chalcedony. Some of the superficial dendroclones are modified: zygomes of adjacent dendroclones are fused, forming a globular knot, 0.10–0.15 mm in diameter, resembling the centrum (i.e. the globular centre) of a spheroclone. As a result, the dendritic character of the zygomes has been lost. Clones between the knots are short and sturdy, 0.25–0.35 mm long, with shafts 0.03–0.05 mm in diameter. Adjacent knots in the distal part of the sponge are often connected with two approximately parallel clones, forming a type of spicule that resembles a didymoclone, a morphology which as yet has only been observed in *Postperissocoelia* n. gen. (discussed above).

Explanation of Plate 10.

Figs 1–3. *Postperissocoelia gnisvardensis* n. gen. et n. sp.

1–3, NRM Sp330; holotype. Erratic from pebble accumulations at the coast near Gnisvärd, Gotland; 1, top view, showing broken heads of the third generation; 2, view of the base, with wrinkled, sub-concentric dermal layer in the centre. In the upper part, basal parts of oscular heads of the second generation are visible; 3, detail of upper part with natural cross-sections of broken oscular heads. Arrows indicate cross-sections of axial canals, into which lateral canals empty.

Figs 4–5. *Postperissocoelia spinosa* (Rigby & Chatterton, 1989)

4–5, Fragment of a small, exceptionally well-preserved specimen from Baillie-Hamilton Island, Arctic Canada, reg. nr. AB 2–99; courtesy of A. Pisera, Warszawa; 4, two oscular centres surrounded by delicately preserved skeleton; 5, detail of Figure 4 demonstrating the skeletal structure, composed of double-connected, didymoclone-like dendroclones. Note the specific feature of small spinose elements.

Scale bars represent 1 cm (Figs 1–4) and 1 mm (Fig. 5).

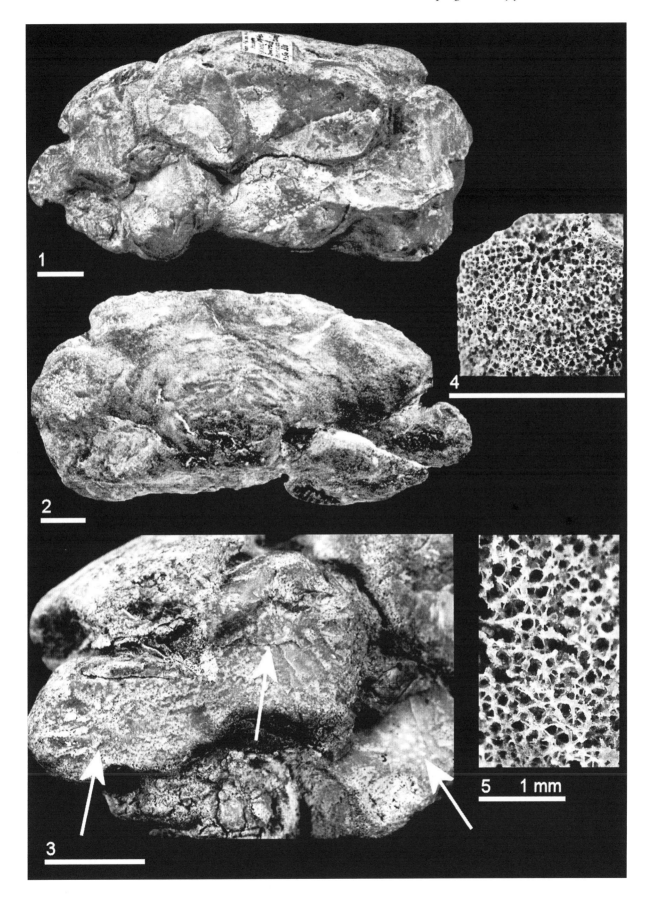

Remarks. – As a unique specimen, care has been taken to confirm the age of this sponge. A small part of the sponge body, as well as a very small quantity of adhering matrix, has been dissolved in hydrofluoric acid and yielded *Nanocyclopia* sp. This long-ranging genus is also known from Upper Ordovician strata (Zwier Smeenk, personal communication 2000), but it was absent from all Gotland samples from Ordovician sponges examined previously.

The specimen is entirely silicified and considerably compressed laterally. Most of the skeletal structure, as well as the system of canals, has been destroyed by chalcedonic replacement that colours most of the sponge interior bluish. In addition, oxidation has coloured most of the silica in the distal zone reddish, identical to that of numerous sponges collected from the same locality. Adhering matrix is greenish-grey, in accordance with those on other sponges. Some monaxons and hexactinellid spicules in adhering matrix also confirm that it is part of the Telychian sponge assemblage. One of the sides shows a natural median section of a series of eight crinoid columnals, the margins of which have been overgrown by the sponge.

The unusual compound growth form may be explained by the genetically defined, limited length of excurrent canals. Canals that achieved their maximum length induced the formation of a new cluster of excurrent canals, in order to drain the newly formed parts. This continuous growth apparently caused the development of new oscular heads, while the formation of stacked skeletal trabs series continued uninterrupted (Rhebergen 2007).

According to Finks (2003), true didymoclones are restricted to the Jurassic and in his opinion, the spicules discussed here should be defined as didymoclone-like dendroclones. They have not yet been observed in Ordovician Orchocladina, but have been described in Silurian anthaspidellid sponges by Rigby & Chatterton (1989), in their description of a Silurian (Wenlock–Ludlow) sponge fauna from Cornwallis Island, Arctic Canada. They erected *Perissocoelia? Spinosa* and figured a knot-like skeletal structure of fused zygomes, as well as the double connections of shafts between the knots (Rigby & Chatterton 1989, p. 29, fig. 5). The characteristic swollen knots and double-connected spicules are shown in a fragment of a small, exceptionally well-preserved specimen from Baillie-Hamilton Island, Arctic Canada, reg. nr. AB 2–99 (Pl. 10, figs 4–5).

Postperissocoelia spinosa (Rigby & Chatterton, 1989) is a compound sponge with numerous oscular outlets, 3–4 mm in diameter, occurring on average 8–10 mm from one another, on both the summit and sides of the sponge body, and containing spinose monaxons in the knots. *P. gnisvardensis* n. sp. has didymoclone-like dendroclones in common with the *P. spinosa*, but differs in its bulbous oscular heads. Spinose monaxons on knots have not been recognized in the new species, but it remains uncertain whether they were originally absent or have been destroyed diagenetically.

Family Chiastoclonellidae Rauff, 1895

Discussion. – Chiastoclonellidae are a relatively small family, compared with related orchocladinid families such as the Anthaspidellidae, Streptosolenidae and Astylospongiidae. According to Rauff (1894), King (1943), Finks (1960), Rigby (1977, 1986), Zhan-Qiu (1981) and van Kempen (1990a), most Chiastoclonellidae are of Permian age and occur predominantly in North America and China. The oldest representatives known to date are erratic, isolated, silicified sponge bodies, of Late Ordovician age, originating from Baltica and collected from fluvial deposits in Germany, the Netherlands, and on Gotland (van Kempen 1990a; Rhebergen & von Hacht 2000; Rhebergen *et al.* 2001).

Many Middle Palaeozoic chiastoclonellid genera, such as the Devonian *Rutkowskiella* Rigby, 1977, *Allossospongia* Rigby, 1986, the Permian *Defordia* King, 1943 and *Actinocoelia* Finks, 1960, contain both tetraclones and chiastoclones, resulting in a pronounced radial and concentric internal organization. This distinguishes them from the Early Palaeozoic chiastoclinellids that occurred from Late Ordovician to Silurian times (*Chiastoclonella* Rauff, 1895 and *Syltispongia* van Kempen, 1990), in which dendroclones form a conspicuous minority (van Kempen 1990a, pp. 152, 155). In reference to this, van Kempen (1990a, p. 159) proposed placing the Devonian and Permian Chiastoclonellidae into a separate group of the Orchocladina, but did not erect a new taxon.

Mehl-Janussen (1999) considered that true tetraclones did not exist until the Mesozoic, with the exception for the Middle Palaeozoic genus *Jereina* Finks, 1960. She regarded the derivation of the chiastoclone from the dendroclone as so decisive that she proposed to refer all orchocladinid taxa with skeletons containing chiastoclones, to the Chiastoclonellidae. Neither Pisera (2002) nor Rigby (2004b) followed this suggestion. Pisera (2002), in his overview of fossil 'lithistids', distinguished two families: the Anthaspidellidae Ulrich *in* Miller, 1889 with predominating dendroclones and the Chiastoclonellidae (Rauff, 1895), with predominating chiastoclones. This has been confirmed in the present study; chiastoclones form an overwhelming majority relative to the questionable occurrence of dendroclones.

Genus *Chiastoclonella* Rauff, 1895

Type species. – *Chiastoclonella headi* Rauff, 1895.

Chiastoclonella sp. Rauff, 1895

Plate 11, figures 1–3

1895 *Chiastoclonella* n. gen. Rauff, pp. 244–247, pl. 17, figs 5–7, pl. 18, fig. 1, figs 98–102.

1990 *Chiastoclonella* Rauff, 1895; van Kempen, pp. 153–156, figs 1a–c, 2a, 3a–b.

2001 *Chiastoclonella* Rauff, 1895; Rhebergen, Eggink, Koops and Rhebergen, p. 98, pl. 22, figs 1–5.

Material. – NRM Sp10106; collected from Gnisvärd.

Description. – NRM Sp10106 is a badly eroded, flattened cushion-like, elliptical sponge, which has been cut axially twice. The massive, acloacate sponge shows irregular horizontal growth increments, sub-parallel to the base. The distal surface is dense and largely featureless. The slightly convex base shows a concentrically wrinkled layer which gradually merges into the sides, where it reaches its maximum diameter (Pl. 11, fig. 1). Remains of an unidentified tabulate occur at the top of the sponge. The reddish-brown surface is smooth, without recognizable pores. The interior consists of bluish chalcedony; the skeleton has been dissolved, except for small pyritized fragments of spicules. Sub-horizontal, slightly undulating darker bands correspond to concentric layers on the exterior, indicating incremental growth stages (Pl. 11, fig. 2). The horizontal layering and the concentrically wrinkled base indicate that the point of skeletal radiation is at the base.

The chiastoclonellid skeleton has been preserved only in a thin, reddish-brown, approximately millimetre-thick basal part. Canals are not preserved, apart from a faint radial pattern. Chiastoclones were the only type of spicules that have been observed. They are 0.05–0.08 mm long, with shafts 0.010–0.020 mm, usually 0.015 mm in diameter (Pl. 11, fig. 3).

Remarks. – Distinguishing chiastoclonellid specimens from rhizoclone-bearing sponges such as *Haplistion* Young & Young, 1877 (see below) was initially difficult, with determinations doubtful due to the poor preservation of the skeleton. The clones of chiastoclones in the former and the tiny strands of rhizoclones in the latter were rendered nearly identical by diagenetic loss of detail. Eventually, a single specimen could be assigned without doubt to *Chiastoclonella*. There may be more specimens, either among the many sponges that have been studied only superficially, or those that are listed as indeterminate Porifera.

The specimen discussed here has been preserved too poorly to assign it to a species, let alone to propose a new taxon, but still certain limitations on its assignment can be drawn. *C. headi* Rauff, 1895 differs by its radiating structure and conspicuously overhanging layers. Late Ordovician specimens of *Chiastoclonella* sp. have been described from Pliocene deposits on the Island of Sylt (Germany) by van Kempen (1990a) and from Lower Pleistocene fluvial deposits in the Netherlands (Rhebergen 1997), but they differ in their sub-spherical body form and central skeletalradiation nucleus.

Syltispongia ingemariae van Kempen, 1990 from the same locations as *Chiastoclonella* sp. is a hemispherical, stromatoporoid-like, laminate sponge with a flat, concentrically wrinkled base. It is composed of a number of sub-parallel, extremely dense, thin layers, with poorly developed canals that empty laterally, as well as isolated, irregularly distributed canals arranged vertically, which empty at the upper surface.

A new, as yet undescribed Ordovician chiastoclonellid species recorded from Lower Pleistocene deposits in the Netherlands, which also occurs in the Ordovician assemblage on Gotland, differs in its cylindrical form, its dense, concentrically wrinkled basal layer and by large, poorly defined gaps distributed over the top and sides.

Subclass Tetractinomorpha Lévi, 1953
Order Streptosclerophorida Dendy, 1924
Suborder Eutaxicladina Rauff, 1894
Family Hindiidae Rauff, 1893

Remarks. – The systematic position of the family Hindiidae Rauff, 1893 is still under discussion. Pisera (2002) placed the family into the suborder Tricranocladina Reid, 1968, in the undefined Palaeozoic lithistids. He regarded the classification of the Hindiidae in Rigby (2004b), which is based on the views by Finks, to be out of date (Andrzej Pisera, personal communication 2006). Mehl-Janussen (1999, pp. 42–43) discussed the taxonomic position of the Hindiidae extensively, also arguing against Finks' opinions. In the present study, however, the classification given in the Treatise of Invertebrate Palaeontology (Finks & Rigby 2004a) has been followed, to maintain unity in the classification used in this study.

Genus *Hindia* Duncan, 1879

Type species. – *Hindia sphaeroidalis* Duncan, 1879.

Hindia sphaeroidalis Duncan, 1879

Plate 11, figures 4–7

1860 *Calamopora fibrosa* n. gen. et sp. Roemer, p. 20, pl. 2, figs 2, 2a–b.

1879 *Hindia sphaeroidalis* Duncan, p. 9, pl. 9, figs 1–6.

1884 *Hindia fibrosa* Hinde, p. 57, pl. 13, figs 1, 1a, b.

1940 *Microspongia sphaeroidalis* Howell, pp. 1–2, pl. 1.

1988 *Hindia sphaeroidalis* Duncan, 1879; Rigby & Webby, p. 61, pl. 26, figs 1–10; pl. 27, figs 1–3.

2004a *Hindia sphaeroidalis* Duncan, 1879; Rigby, pp. 80–82.

Material. – NRM Sp10108, NRM Sp10110, NRM Sp10118, NRM Sp 10140, NRM Sp10145, NRM Sp10146, NRM Sp10152, NRM Sp10171 and NRM Sp10199; collected from Blåhäll/Tofta. NRM Sp10197; from Gnisvärd.

Other material. – NRM Sp10064, NRM Sp10096 and NRM Sp10126; from Blåhäll/Tofta. NRM Sp10095; from Gnisvärd.

Description. – *Hindia sphaeroidalis* is a spherical to sub-spherical sponge with straight, radially arranged in- and excurrent canals. Sub-spherical specimens vary from 10 to 55 mm in diameter, but most are 25–30 mm across (Pl. 11, fig. 4). The largest specimen (NRM Sp10197) is an abraded partial sponge, 55 mm long. Specimens broken medially show the strictly radial structure of stacked spicules and canals (Pl. 11, fig. 5). Usually, the regular distal growth is reflected in the formation of concentric layers (as in Mehl-Janussen 1999).

Incurrent and excurrent canals are very similar. They are arranged radially, paralleling the skeletal strands. Canals of five specimens have been measured. The largest canals, presumed to be excurrent, are 0.25–0.30 mm across. The smallest, probably incurrent, canals, are 0.10–0.12 mm across. Most canals are of intermediate size, 0.15–0.20 mm in diameter. Intercalation of new canals occurs as surface area increases with the diameter of the sponge during growth. The surface is generally smooth, showing cross-sections of canals surrounded by small triangles, which are brachyomes in cross-section (Pl. 11, fig. 6).

The skeleton of *H. sphaeroidalis* is composed of tricranoclone desmas. A tricranoclone has three cladomes, each of which has sculptured brachyomes (Fig. 6). The cladomes are connected with those of adjacent spicules, always at an angle of 120°, in such an array that a brachyome wraps around an adjacent arched cladome at its 'shoulder'. Series of spicules in alternating distal and proximal arrangement form a zig-zag structure between adjacent canals (Pl. 11, fig. 5). Tricranoclones could be measured only rarely, as most have been diagenetically modified. They are 0.20–0.30 mm long, measured from ray-tip to ray-tip. Rays are 0.15–0.20 mm long and 0.05–0.08 mm in diameter.

The central parts of *H. sphaeroidalis* are often hollow; this probably occurred by resorption of silica of the oldest spicules during life. Specimens exhibiting this phenomenon have their centre infilled by chalcedony.

Remarks. – *Hindia sphaeroidalis* is a common sponge species from North America, Australia and Europe. Its cosmopolitan character and its long biostratigraphical range (Middle Ordovician to Devo-

---

Explanation of Plate 11.
Figs 1–3. *Chiastoclonella* sp.
1–3, NRM Sp10106; **1**, view of concentrically wrinkled base; **2**, longitudinal plane showing growth increments, corresponding with those in Figure 1; **3**, detail of Figure 2, showing skeletal structure of irregularly arranged chiastoclones in the basal part.
Figs 4–7. *Hindia sphaeroidalis* Duncan, 1879
4, NRM Sp10171; cross-section, showing the strictly radial pattern of incurrent and excurrent canals. **5**, NRM Sp10145; natural longitudinal section showing the radial canals. **6**, NRM Sp10118; view on distal surface in which circular cross-sections of canals are surrounded by triangular cross-sections of tricranoclones. The grey patch is the remains of the encrusting sponge *Opetionella incompta* n. sp.. **7**, NRM Sp10152. Two specimens of *Favosites* sp. (see arrows) that have nucleated on the sub-spherical sponge body.
Scale bars represent 1 cm (Figs **1, 2, 4–7**) and 1 mm (Fig. **3**).

nian) are unusual, compared with related taxa. Rigby & Chatterton (1989, p. 35; 1999, p. 18) assumed *Hindia* to have been planktonic or rolling on the sea-floor, but de Freitas (1991, p. 2051) considered *Hindia* to have been benthic, on account of its abundant occurrences in sponge biostromes of platform slope sequences, and its dense and rigid skeletal structure.

Most of the specimens collected from Gotland are probably part of the Ordovician assemblage. The great majority, however, lack any adhering matrix to serve to separate them easily, and this is also true of specimens in the Telychian assemblage. In the Silurian association *Hindia* is the third most frequent genus, comprising about 200 specimens, which is about 10 per cent of the assemblage. The Telychian age of around ten specimens is proven through co-occurring fossils, adhering monaxons and hexactinellid spicules, as well as through acritarchs obtained from dissolved matrix samples. The remaining majority of specimens could be distinguished from Ordovician specimens mainly on account of their reddish colouring, differences in the adhering matrix (where present), absence of isolated spicules in Ordovician matrix, and by coarser chalcedony in the Ordovician material.

In NRM Sp10199, *H. sphaeroidalis* occurs with *Caryoconus gothlandicus* (Schlüter, 1884), and in NRM Sp10108 with *P. subrectus*. Some areas of the surface of NRM Sp 10118 are covered with abraded remains of *O. incompta* sp. nov. (Pl. 11, fig. 6). Allochthonous monaxons and hexactinellid spicules confirmed the age of two specimens: NRM Sp10198 and NRM Sp10152. Some specimens have been partly overgrown by the tabulate *Favosites* sp. (Pl. 11, fig. 7), as well as by *Catenipora* sp (NRM Sp10146), which considerably impacted on the regularity of the growth of the sponge.

Suborder Sphaerocladina Schrammen, 1910
Family Astylospongiidae Zittel, 1877

Remarks. – Genera in the Astylospongiidae are distinguished by the arrangement of canals and exhalant pores (Finks & Rigby 2004a). They show a wide range of body shapes. Ordovician representatives are generally globular, and only rarely obconical or pyramidal. Silurian astylospongiids show more variation in body shape, a feature which increases in Devonian and Permian genera, including cylindrical, tubular and vasiform growth forms. The skeleton is composed of spheroclones.

Genus *Caryoconus* Rhebergen & van Kempen, 2002

Type species. – *Astylospongia gothlandica* Schlüter, 1884.

Caryoconus gothlandicus (Schlüter, 1884)

Plate 12, figures 1–8

1893 *Caryospongia diadema* (Klöden, 1834); Rauff, p. 304, pl. 10, fig. 6.

1983 *Caryospongia diadema* (Klöden, 1834); van Kempen, pp. 100–104, figs a–k.

2000 'Astylospongia' gothlandica Schlüter, 1884; Rhebergen & von Hacht, p. 343, fig. 12.

2002 *Caryoconus gothlandicus* (Schlüter, 1884); Rhebergen & van Kempen, pp. 188–192 (*cum syn.*), figs 3–13.

Emended diagnosis. – Irregularly stalked, sub-spherical to obconical astylospongiid, generally with one apical oscular centre; sub-spherical and stalked parts with irregular, anastomosing or branched, surficial furrows that radiate from bulge situated at maximum diameter; furrows commonly extending along stalk. Numerous small prosopores and fewer larger apopores distributed irregularly over surface. More-or-less straight canals radiating from centre of sub-spherical part of sponge body. Stalk with conspicuous interior cone of straight, diverging canals, extending from centre to end of the stalk. Straight,

---

Explanation of Plate 12.
Figs 1–8. *Caryoconus gothlandicus* (Schlüter, 1884)
1–2, NRM Sp10111; 1, lateral view. The short stalk is situated opposite the oscular centre; 2, lateral view, showing *Pentameroides subrectus* attached to the lateral surface of the sponge body. 3, NRM Sp10161; lateral view of a slender form. Grooves converge to the oscular centre. Most of the numerous apopores empty into grooves. 4, NRM Sp10116; lateral view. Deep lobate grooves converge to the oscular centre at the left. 5–6, NRM Sp10138. Axial plane, demonstrating form and height of the cone, radiating canals in the sub-spherical upper part, and sub-parallel canals, more-or-less perpendicular to both cone and external surface. 7, NRM Sp129. Axial plane of a sub-spherical specimen with well-preserved skeleton, canal system and short cone. 8, NRM Sp10158. Lateral view of a specimen with the oscular centre at the top, the cone (not visible) to the left, and rare outgrowths at the base.
Scale bars represent 1 cm.

parallel canals, perpendicular to sides of cone, running towards surface. Principal choanosomal skeleton composed of interconnected regular spheroclones, but spheroclones within the cone often irregular, with longitudinally elongated rays, forming diverging strands (emended from Rhebergen & van Kempen 2002).

Material. – NRM Sp129, NRM Sp10065, NRM Sp10066, NRM Sp10067, NRM Sp10068 and NRM Sp10111; collected from Gnisvärd. NRM Sp10116, NRM Sp10138, NRM Sp10139, NRM Sp10154, NRM Sp10158, NRM Sp10161 and NRM Sp10265; from Blåhäll/Tofta and NRM Sp4511; from an unknown location on Gotland.

Other material. – NRM Sp10160, NRM Sp10162, NRM Sp10163, NRM Sp10177, NRM Sp10230, NRM Sp10231, NRM Sp10232, NRM Sp10233, NRM Sp10234, NRM Sp10235, NRM Sp10236, NRM Sp10237, NRM Sp10238, NRM Sp10239, NRM Sp10240, NRM Sp10241, NRM Sp10242 and NRM Sp10243; all from Blåhäll/Tofta.

Description. – *Caryoconus gothlandicus* is an asymmetrical astylospongiid sponge with one (rarely two) apical oscular centres and with a wide range of body shapes, varying from obconical to sub-spherical. All forms have a sub-globular part that merges into a stalked, sub-conical part (Pl. 12, figs 1–7). The sub-globular part has a bulge on one side that ends into a convex apical expansion, which always has a smooth surface. Several furrows, 1–3 mm wide and 0.7–2.5 mm deep, radiate irregularly from the sides of the expansion and over the surface, running sub-horizontally or tangentially downwards towards the stalked part (Pl. 12, figs 1, 3–4). These furrows branch and anastomose frequently, but generally do not meet on the opposite lateral side of the sponge body. Usually, the distal parts of the furrows tend to bend towards the stalked part. Elongate specimens, such as NRM Sp10161, have a smooth surface opposite the apical expansion. Some specimens are slender, more sub-cylindrical than pyriform, with length up to twice their maximum width, and often showing a gradual change from the upper to the lower part (Pl. 12, figs 3–6). In general, their apical expansion is less prominent and the sides with the expansion are commonly convex, whereas the opposite side is straight or slightly concave. In some specimens, the stalked part and consequently also the cone are conspiculously bent, as shown in NRM Sp10138 (Pl. 12, figs 5–6).

The collection also includes sub-globular specimens such as NRM Sp129, in which the diameter approximately equals the length (Pl. 12, figs 1, 7). Here, the stalk is poorly developed and the skeletal cone is much shorter than usual. Some specimens have two oscular centres on the presumed upper surface (see below). Despite the differences in body shape, the canal system is in general structure similar in all specimens.

The stalked part is unusual and differs from the other parts of the sponge both in skeletal composition and canal system. Independently from the body shape, each specimen has a characteristic, sharply angled skeletal cone, unique among Astylospongiidae, but as yet its function remains enigmatic. The top of the cone reaches from its base up to 1/2 or 3/4 of the height of the sponge and has its maximum width at the top of the stalked part (Pl. 12, fig. 5–6). The angle of the cone ranges from 20 to 30°. The sides of the cone are straight or slightly curving.

The canal system differs from those known in other Astylospongiidae, consisting of three series of more-or-less straight canals, 0.3–0.5 mm wide. Figure 8 shows a schematic reconstruction. One series in the sub-spherical part of the sponge body radiates from a centre at the top of the cone, towards the outer surface. The second series, in the stalked part of the sponge, consists of straight, parallel canals, running from the sides of the cone towards the outer surface. The third series form the cone itself: straight, diverging, sub-longitudinal canals extending from the top towards its base. A connection between the parallel canals and those in the cone could not be established. Prospopores and apopores are distributed evenly over the surface, but excurrent canals, 0.5 mm in diameter, empty preferentially into the furrows (Pl. 12, fig. 3) with associated pores 0.5–0.75 mm in diameter.

The choanosomal skeleton is composed of a regular mesh of connected spheroclones, as shown in Figure 9. Spheroclones are 0.20–0.27 mm in diameter, but usually 0.23 mm in diameter (Pl. 12, fig. 7). In contrast, spheroclones in the cone are irregular in size, tending to combine to form diverging strands, with some of their rays elongated in the main direction of the cone, as shown in Figure 10. They are 0.16 mm measured along the short rays to 0.33 mm along the long rays. Some spheroclones have elongated rays at right angles to the short rays, connecting centra of spicules that form the walls of a canal. The centre of skeletal growth is at the top of the cone.

Remarks. – Co-occurrence of *Caryoconus* with *P. subrectus* Schuchert & Cooper, 1931 is considerably more frequent than it is with anthaspidellids. One of best kept specimens of *P. subrectus* is connected directly to the surface of the sponge body

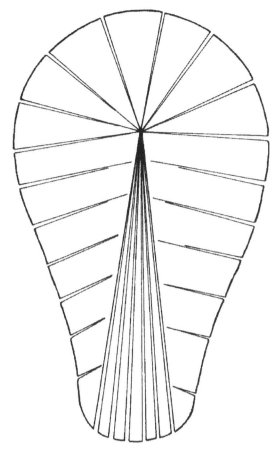

Fig. 8. Schematic reconstruction of the unusual sphaeroclonid *Caryoconus gothlandicus*, the most frequently occurring species of lithistid in the assemblage.

(Pl. 12, fig. 2). NRM Sp10265 is the only specimen that is embedded in greyish-green matrix, comprising several specimens of *Pentameroides*. Part of the matrix has been dissolved in hydrofluoric acid, but the sample did not contain acritarchs.

The orientation and possible living position of *Caryoconus* has been discussed by Rhebergen & van Kempen (2002). Their discussion excluded an upright position, with the stalked part downwards and, consequently, the open side of the cone as the base. They instead interpreted a sub-horizontal position of the sponge during life, with the lateral bulge attached to the substrate, and the cone representing a series of excurrent canals. In that position, canals in the cone point obliquely upwards and could have emptied easily, especially in those specimens that show a concavely bent stalk region.

The availability of abundant new material (together with discussions with Andrzej Pisera) has led to a re-interpretation of the life position of the sponge. We now regard *Caryoconus* as having indeed been oriented sub-horizontally, but axially reversed from the previous interpretation. In this view, the lateral grooves led to the bulge, which is regarded as the primary exhalent outlet or oscular centre (Pl. 12, fig. 3, top and Pl. 12, fig. 4, left). This arrangement is similar to that seen in the closely related *Caryospongia* species, such as *C. juglans* (Quenstedt, 1878), *C. diadema* (Klöden, 1834) and *C. edita* (Klöden, 1834). We interpret the cone of *Caryoconus* as a specialized structure containing numerous incurrent canals. An aberrant growth form seen in some recently collected specimens supports this view: some specimens developed simple outgrowths, opposite the oscular centre (Pl. 12, fig. 8). However, in some specimens, the apical expansion and the end of the conical structure are opposite, but no lateral attachment site to the substrate can be recognized (NRM Sp10111; Pl. 12, fig. 1). Thus, the life position and the function of the conical construction remain partly problematic and need further study.

The Leipnitz Collection contains more than 1000 specimens of *C. gothlandicus* collected from the Gnisvard/Tofta/Blåhäll area during the past decade, making this species the most abundant in the fauna; it comprises about 50 per cent of the total assemblage. A similar ratio is characteristic of the de Vries Collection, which comprises ca. 800 sponges. This could be argued to indicate an endemic character for the sponge association as a whole, despite the similarity of many rare taxa to species known from Arctic Canada.

Rigby & Chatterton (1999) described and figured two specimens of a Silurian (Wenlock) astylospongiid, *Astylospongia*? *praemorsa* (Goldfuss, 1826) from the Mackenzie Mountains, northwestern Canada. It has the stalked body form with a subglobular and a stalked part in common with *C. gothlandicus*, but its excurrent canals are concentrically arranged, parallel to the surface, and empty into a conspicuous spongocoel. This array is more similar to *Astylospongia* Roemer, 1860 than to *Caryoconus* Rhebergen & van Kempen, 2002. In addition, an internal conical skeletal structure in the stalked part has not been reported.

Genus *Lindstroemispongia* n. gen.

Type species. – *Lindstroemispongia cylindrata* n. sp.

Derivation of name. – Sponge named in honour of the Swedish palaeontologist Gustaf Lindström (1829–1901), who probably marked the holotype with his initials.

Diagnosis. – Erect, cylindrical sponge with rounded top, but base unknown. Spongocoel or oscular area

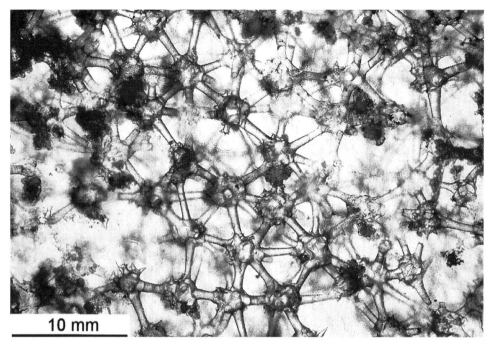

Fig. 9. Regular choanosomal skeleton in *Caryoconus gothlandicus* composed of spheroclones preserved in clear chalcedony. NRM Sp129 (photo: T. M. G. van Kempen).

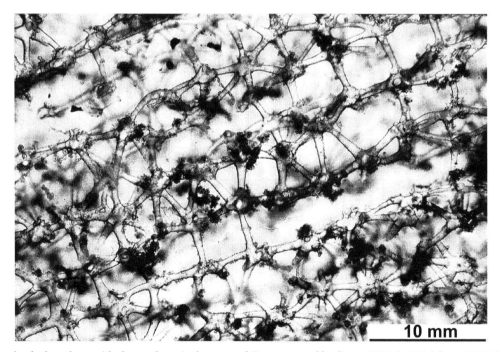

Fig. 10. Strands of spheroclones with elongated rays in the conus of *Caryoconus gothlandicaus*. NRM Sp4511 (photo: T. M. G. van Kempen).

absent. Outer surface covered with rosettes, each ca. 25 mm in diameter, formed by clusters of radiating furrows, running over the surface. Cluster of apopores in poorly defined centres of rosettes. Apopores usually empty into proximal parts of surficial furrows. Numerous radially arranged, vertically stacked series of probably exhalant canals flare upwards and outwards from the axial plane, bending convexly to nearly 90° at the surface. Numerous inhalant pores scattered evenly over the outer surface. Arrangement of inhalant canals unknown. Skeleton composed of spheroclones.

Remarks. – Exhalant canals in the Astylospongiidae are arranged either radially, such as in *Carpospongia* Rauff, 1893 and *Caryospongia* Rauff, 1893, are regularly concentric as in *Astylospongia* Roemer, 1860, *Palaeomanon* Roemer, 1860 and *Astylospongiella* Rigby & Lenz, 1978, or irregularly concentric and anastomosing, as in *Tympanospongia* Rhebergen, 2004. The Lower Silurian *Caryoconus* Rhebergen & van Kempen, 2002 has a distinct arrangement of excurrent canals: radial in the upper part, sub-horizontal in the lower part, and with a cone of diverging canals normal to the parallel canals (discussed above). Both body form and canal system of *Lindstroemispongia* n. gen. are fundamentally different.

Rauff (1893) and von Hacht (1990) figured surficial rosettes in the globular *Carpospongia stellatimsulcatum* (Roemer, 1848) showing poorly defined centres of furrows distributed over the surface. Rauff (1893) also described and figured the only specimen known at that time of *Carpospongia conwentzi*, a globular sponge with surficial tumuli (Rauff 1893). The tumuli are usually abraded into an elliptical teardrop shape, the tapering ends pointing to a slightly lowered centre, and separated by poorly defined radial grooves. The exterior of the sponge body is, as a result, marked by a polygonal pattern, the facets of which show virtual rosettes (Rhebergen *et al.* 2001). However, the canal system of *C. conwentzi* is strictly radial, in contrast to the more complex arrangement in *Lindstroemispongia* n. gen.

Caliculospongia Foerste, 1916 is a small cylindrical sponge from the Upper Ordovician of Kentucky (USA). It differs from *Lindstroemispongia* n. gen. in having a distinct cloacal depression at the apex, surrounded by a more-or-less flat rim.

Astylospongiid genera with a cylindrical body form are more characteristic of the Devonian and Permian, for example *Astylotuba* Rigby *et al.*, 2001 from the Devonian of the Holy Cross Mountains, Poland. This species possessed a distinctive diplorhizal canal system in craticulariid pattern, and there is no close similarity to the new Silurian taxon. *Attungaia* Pickett, 1969, from the Devonian of Australia, differs dramatically in being tubular and branching; it also shows a thin-walled construction, each branch with a deep spongocoel.

Camellaspongia Rigby & Bayer, 1971 and *Phialaspongia* Rigby & Bayer, 1971, both from Upper Ordovician strata of Minnesota (USA), are both conical and differ from *Lindstroemispongia* n. gen. in having a deep exhalant depression into which excurrent canals empty.

Inglispongia Pickett, 1969 is a large, massive sponge, cylindrical, flabellate or irregular in body form, and lacking a cloaca, but having large, parallel, vertical exhalant canals in the axial region, discharging onto the surface. *Lindstroemispongia* n. gen. differs in having exhalant canals opening into the grooves of the stellate rosettes.

Other astylospongiid genera differ even more than those cited above, being flabellate, laminate, obloid to biscuit-shaped, and with canal systems that are fundamentally distinct from those of *Lindstroemispongia* n. gen.

Lindstroemispongia cylindrata n. sp.

Plate 13, figures 1–7

Derivation of name. – Describing the cylindrical form of the sponge body.

Diagnosis. – As for the genus.

Material. – LMG–G1886, the holotype and only specimen known to date, is an erratic from Visby, Gotland (Sweden), and is housed in the collections of the County Museum of Gotland, Visby (Sweden). The specimen bears an ink-written indication reading 'Wisby' and 'G.L.', short for Gustaf Lindström (Pl. 13, fig. 1). It has been cut twice: transversely, probably by Gustaf Lindström, and also, for this work, axially at the Vrije Universiteit Amsterdam; both surfaces have been polished. One of the halves has been glued. The transverse polished surface was covered with a thick layer of old lacquer, which had to be removed for the present examination.

Description. – The sponge body is incomplete, and a section of unknown length is missing. The preserved part is a regular cylinder, 66 mm high and 30 mm across (Pl. 13, figs 1–2). The top is sub-hemispherical and has a smooth surface. The basal part is missing. Lateral surfaces show rosettes, 20–25 mm in diameter, formed by radiating furrows (Pl. 13, figs 2–3); ten to twelve irregular furrows, 6–13 mm long, 1 mm deep and 0.8 mm wide, radiate from some poorly defined centres along the outer surface. Furrows of adjacent rosettes are slightly connected to each other. Some of the furrows bifurcate distally. Rosettes have been preserved on one side of the cylinder only. The opposite side is abraded, but vague remains of furrows indicate that the entire lateral surface was also originally covered by rosettes (Pl. 13, fig. 1). Inhalant pores are presumed to have been present over the entire surface, but could not be observed due to poor preservation.

The polished surfaces of the cross-section (Pl. 13, fig. 4) show a regular pattern of radiating canals. The median section of the cylinder (Pl. 13, fig. 5) shows the pattern of diverging canals, which are on average 0.20 mm in diameter. In the lower and middle part, they flare upwards and outwards from a virtual central axis and join the surface at an angle of ca. 90°. Larger canals, with a diameter of 0.25–0.30 mm, radiate from the centre to the outer surface, where most empty into or near the furrows, close to the centres of the rosettes. The upper part of the cylinder contains canals that are initially arranged vertically along the virtual central axis, but at around 10 mm from the outer surface, they curve smoothly, until they reach the surface at an angle of about 120°. In the hemispherical apex, a radial pattern of divergent canals is visible, starting in the centre at about 14 mm from the summit.

The skeleton is composed of spheroclones, but is poorly preserved. Spheroclones are best-preserved at the surface (Pl. 13, fig. 6). Internally, most have been diagenetically dissolved, but a few are recognizable (Pl. 13, fig. 7). Spicules are 0.22–0.27 mm across (mean 0.23 mm). Globular centra of fused clones are 0.05 mm across, and individual clones are 0.03 mm across. The position of the initial skeletal nucleation point could not be established, but was presumably situated in the missing basal part.

Remarks. – The characteristic surficial rosettes consist of grooves that presumably led water from the apopores to a central exhalent pore within unpreserved organic tissue. This organization is similar to that in some Ordovician globular species of *Caryospongia* Rauff, 1893. Considering that rosettes in *L. cylindrata* appear over the entire cylindrical surface, the living position of this sponge was most probably erect.

Lindstroemispongia cylindrata n. gen. et n. sp. is a unique specimen, which was recently recognized in the collections of the County Museum of Gotland, Visby, where it had remained unnoticed, probably for more than a century. Because of the absence of adhering sediment, its stratigraphical position cannot be obtained by acritarchs. However, it has most of the features in common with other species of the present assemblage (see Introduction). Fragments of allochtoneous monaxons on the surface are also an indication of its Telychian age. Although there is only circumstantial evidence, assignment to the Telychian assemblage appears sound.

Order Spirosclerophorida Reid, 1963
Suborder Rhizomorina Zittel, 1878a

Remarks. – The rhizomorine taxa described below constitute the first recorded examples from Baltica of the Palaeozoic Suborder Rhizomorina.

The suborder Rhizomorina was proposed by Zittel (1878) to include sponges with a skeleton composed predominantly of rhizoclone desmas and entirely monaxonid spicules. Rigby & Webby (1988, p. 16) placed them into the order Lithistida Zittel, 1878a, but Finks & Rigby (2004a) referred them to the order Spirosclerophorida Reid, 1963 in the subclass Lithistida Schmidt, 1870.

Family Haplistiidae de Laubenfels, 1955

Remarks. – Rigby & Webby (1988) synonymized the Columellaespongiidae Pickett, 1969 with the Haplistiidae de Laubenfels, 1955 and demonstrated the occurrence of Haplistiidae as early as in the Late Ordovician. Rigby & Chatterton (1989) stated that the family had a greater diversity in the Lower Palaeozoic than in the Middle Palaeozoic. The geographical range includes Australia, North America and Eurasia.

Genus *Haplistion* Young & Young, 1877

Type species. – *Haplistion armstrongi* Young & Young, 1877.

Remarks. – Finks (1960, p. 57) summarized useful characters for differentiation of the species in *Haplis-*

Explanation of Plate 13.
Figs 1–7. *Lindstroemispongia cylindrata* n. gen. et n. sp.
1–7, LMG G1886; holotype. Erratic from Norderstrand, Visby (Gotland, Sweden); **1**, lateral surface, with handwritten 'Wisby' and initials 'G L'. Surface abraded, but remains of rosettes are recognizable; **2**, opposite lateral surface with conspicuous rosettes; **3**, detail of lateral surface. Each rosette has a central cluster of radial apopores and is formed by converging surficial grooves, into which radial canals empty; **4**, cross-section, showing the pattern of radiating canals; **5**, median view of polished surface. Radial canals are arranged pinnately, flaring upwards and outwards to meet the surface at about 90°; **6**, spheroclones, visible on some parts of the outer surface; **7**, a few spheroclones recognizable in the distal part of the interior.
Scale bars represent 1 cm (Figs **1–6**) and 1 mm (Fig. **7**).

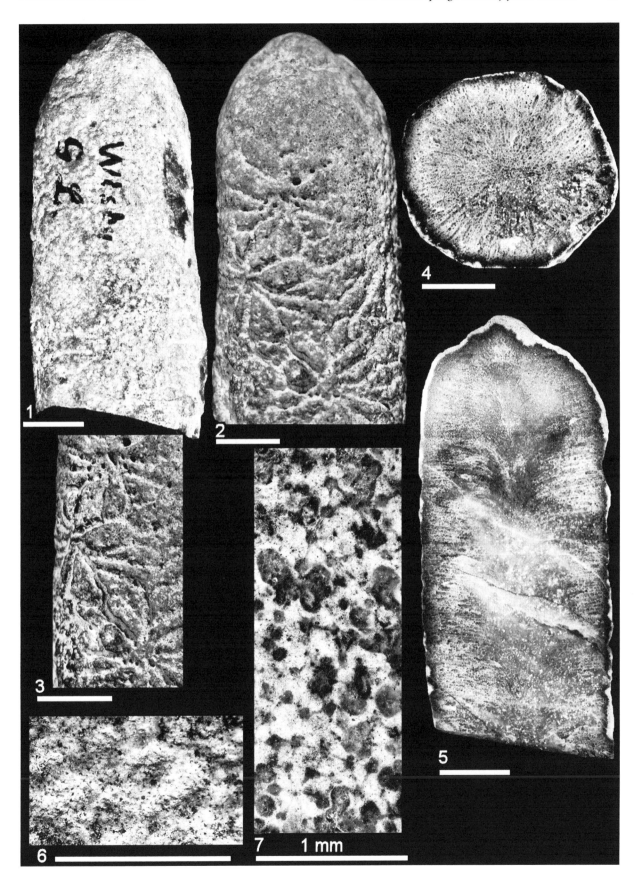

tion as follows: (1) body form, (2) dimensions of mesh spaces, (3) thickness of the spicule tracts and (4) form of larger canals (if present). He considered the diameter of spicule tracts to be the most important, and dimensions of mesh spaces second most important. This set of characters was applied to the 15 valid species of *Haplistion* known at that time, producing the following groupings (Finks 1960, p. 88, summarized herein):

Group 1: Mesh spaces less than 1 mm; radial and horizontal tracts equal in diameter.

Subgroup 1A: tracts not exceeding 0.5 mm in diameter.
Subgroup 1B: tracts about 0.5 mm in diameter.

Group 2: Mesh spaces 1.0 mm or more; horizontal tracts half the diameter of radial tracts.

Subgroup 2A: radial tracts 0.5–1.0 mm in diameter.
Subgroup 2B: radial tracts 1.0–2.0 mm in diameter.

Finks (1960), Rigby & Dixon (1979) and Rigby & Webby (1988) have discussed the possible occurrence of true dendroclones. Finks (1960, p. 89) emphasized that 'where spicules occur singly rather than in bundles (…) the amphiarborescent processes are developed only at the end of the shaft, and the spicules are indistinguishable from an anthaspidellid dendroclone'. He considered *Haplistion* to be both related to the orchocladinid Anthaspidellidae and on the other hand to be a true rhizomorine sponge, as it contains desmas characteristic of both groups. Rigby & Dixon (1979, p. 599) stated that in *Haplistion minutum* 'dendroclones are moderately common spicules', but Rigby & Webby (1988, p. 16), in their diagnosis of *Haplistion regularis* noted: 'common pin-like single dendroclones (?) also cross-connect horizontally' and in the description: 'Isolated horizontal dendroclones (?) have pin-like shafts' (Rigby & Webby 1988, p. 16). In the present study, these desmas are called 'dendroclones', as the material is insufficiently well preserved to allow a resolution to the problem.

Haplistion minutum Rigby & Dixon, 1979

Plate 14, figures 1–6

1979 *Haplistion minutum* Rigby & Dixon, pp. 598–602, figs 5–7.

Material. – NRM Sp10125; collected from Gnisvärd. NRM Sp10170, NRM Sp10172 and NRM Sp10178; from Blåhäll/Tofta.

Other material. – NRM Sp10244, NRM Sp10245, NRM Sp10246 and NRM Sp10250; from Blåhäll/ Tofta and NRM Sp10269; from Gnisvärd. NRM Sp862 and NRM Sp863 are from unknown localities on Gotland. NRM Sp10190, NRM Sp0256, NRM Sp10266 and NRM Sp10267, from Blåhäll/Tofta, are assigned to *Haplistion* sp. cf. *H. minutum.*

Description. – *Haplistion minutum* is a massive, acloacate sponge with a very fine skeletal net of radiating or ascending columns, which are cross-connected by minor concentric or horizontal tracts. Horizontal or concentric tracts are spaced more irregularly than the vertical series.

Fifteen abraded specimens have been identified as *H. minutum*. All specimens are massive but flattened sponges. Surfaces are smooth, sometimes with a ridge or shallow groove, where washed-in monaxons and hexactinellid spicules are present in adhering matrix. Between one and three small dimples, each about 4 mm across and up to 2 mm deep, occur on most of the larger specimens, but their significance is unknown. Eight of the specimens have been cut medially to reveal the growth nucleation point, which is at the flattened base (see below).

The largest specimen, NRM Sp10125, is 52 × 34 mm and 60 mm high. In longitudinal view, it is triangular, with a flattened base (Pl. 14, fig. 1). The sponge larva apparently settled onto a solid substrate; this feature is also seen in several other specimens. In NRM Sp10178, a single brachiopod valve is incorporated into the sponge body (Pl. 14, fig. 2).

The interior of NRM Sp10170 consists of bluish, translucent chalcedony, in which most of the skele-

---

Explanation of Plate 14.
Figs 1–6. *Haplistion minutum* Rigby & Dixon, 1979
1, NRM Sp10125. Polished median section, demonstrating the grid of radial tracts, which are cross-connected by concentric tracts. Growth initiated on an incorporated organism, which has been converted into agate-like chalcedony. **2**, NRM Sp10178. Median section, showing the same pattern as in Figure 1, but sub-horizontal, concentric tracts are more conspicuous. The sponge larva settled on a brachiopod. **3, 5–6**, NRM Sp10170; **3**, polished median section, showing the structure composed of radiating tracts; **5**, detail of the counterpart of Figure 3, showing well-preserved radiating tracts; **6**, detail of Figure 5, showing racts composed of tiny rhizoclones, bridge-connected by amphiarborescent dendroclones. **4**, NRM Sp 10172. Oblique view of natural surface with concentric tracts cut obliquely. Scale bars represent 1 cm (Figs **1–5**) and 1 mm (Fig. **6**).

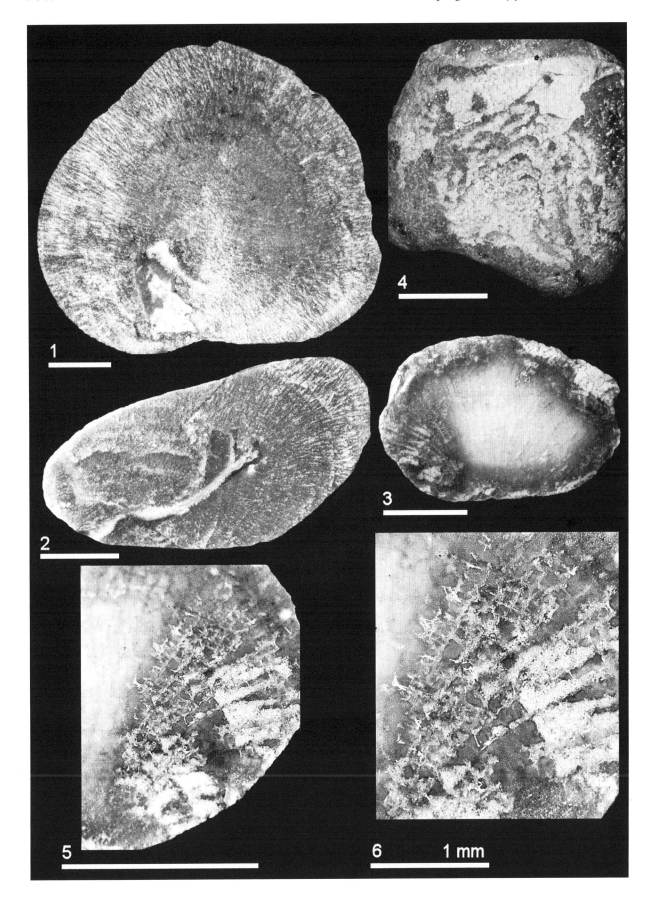

ton has been dissolved, as shown in Plate 14, figure 3. Part of the flattened top is covered by matrix. The axial plane shows a pattern of radial canals. They are straight in the central section, but are convexly curved laterally, upward and outward, to meet the surface at approximately right angles. The concentric canal system is less developed or has been preserved poorly. The two canal systems are therefore disposed perpendicularly to each other at the surface and form a dense grid. NRM Sp10172 is a small, sub-cylindrical sponge with small obconical base. Most of the skeletal structure has been lost through chalcedony replacement, but a near-surface breakage plane shows spicule tracts in a natural oblique section (Pl. 14, fig. 4).

Radiating canals are 0.20–0.35 mm in diameter and are separated by 0.6–0.8 mm. Sub-horizontal, concentrically arranged canals cross-connect the radiating tracts perpendicularly, but they are poorly preserved in the available specimens. The few measurable concentric canals are 0.25–0.50 mm in diameter.

The skeleton consists of tracts composed of bundles of sub-parallel rhizoclones. Tracts are arranged three-dimensionally to enclose spaces in the dermal mesh, which form the walls of canals (Pl. 14, fig. 5). Mesh spaces are 0.10–0.25 in diameter. Tract dimensions are consistent for a particular position in the sponge, measuring 0.03–0.05 mm in diameter in the interior of the basal part of NRM Sp10170, and a maximum of 0.10 mm (mean 0.04 mm) on the surface of NRM Sp10172. These dimensions fall into Finks' Group 1A.

In the basal part of NRM Sp10170, horizontally arranged dendroclones occur in radiating canals as bridges between two adjacent vertical tracts. These dendroclones are amphiarborescent, 0.15–0.35 mm long, with shafts 0.015–0.030 in diameter (Pl. 14, fig. 6). Their zygomes are connected with those of adjacent vertically arranged rhizoclones. Monaxons appear to be absent.

Remarks. – Most related species are easily separated from *H. minutum*. Tracts in *Haplistion armstrongi* Young & Young, 1877 contain many more rhizoclones, whereas *Haplistion cylindricum* Rigby & Dixon, 1979 is sub-cylindrical with marked annuli, nodes and a deep, broad spongocoel. *Haplistion toftanum* n. sp. is cylindrical, with a conspicuous canal system of axial, converging and diverging canals.

Skeletal details in the two known specimens of *Haplistion creswelli* Rigby & Dixon, 1979 are poorly preserved, due to replacement by crystalline calcite, which makes comparison difficult. However, cross-connecting dendroclones are absent, providing at least one feature separating it from *H. minutum*.

The Devonian *Haplistion aeluroglossa* Finks, 1960 differs from *H. minutum* in its larger rhizoclones (0.3–0.8 mm long, 0.05 mm in diameter), its rectangular radial canals, the occurrence of oxeas and the distinct body shape: flattened ovoid to lobate or sub-digitate.

Haplistion regularis Rigby & Webby, 1988 is similar to *H. minutum* in many respects, such as body form, dimensions of rhizoclones, diameter of tracts and the presence of cross-connecting dendroclones between dominantly radial tracts, and it is therefore assignable to Finks' Group 1. However, radial tracts in *H. regularis* are twice the diameter of horizontal tracts. In that respect, it belongs in Finks' Group 2 and is distinct from *H. minutum*.

Haplistion cylindricum Rigby & Dixon, 1979

Plate 15, figures 1–4

1979 *Haplistion cylindricum* Rigby & Dixon, pp. 592–598, fig. 4, pl. 2, fig. 4.

Material. – NRM Sp10156 collected from Blåhäll/Tofta.

Other material. – NRM Sp947 from Stengård, Gnisvärd.

Description. – *Haplistion cylindricum* is a tall, sub-cylindrical sponge with a broad, open and deep spongocoel, extending to the base. The skeleton is

---

Explanation of Plate 15.
Figs 1–4. *Haplistion cylindricum* Rigby & Dixon, 1979
1–4, NRM Sp10156; **1–2**, lateral dermal surface. 'F' indicates *Favosites* sp.; **3–4**, Axial view showing the deep, open spongocoel. The sponge wall contains both vertical canals, bending outwards to meet the dermal surface, and sub-horizontal radial canals, emptying at the gastral surface. Irregular projections from the gastral surface mark the apopores of radial canals, emptying into the spongocoel.
Figs 5–6. *Haplistion toftanum* n. sp.
5–6, NRM Sp10157; holotype. Erratic from gravel accumulations at the coast at Blåhäll/Tofta (Gotland, Sweden); **5**, lateral surface with numerous randomly distributed apopores; **6**, axial plane with a shallow spongocoel. Both large axial canals and sub-radial canals, converging from the distal part to the axial region, are apparent.
All scale bars represent 1 cm.

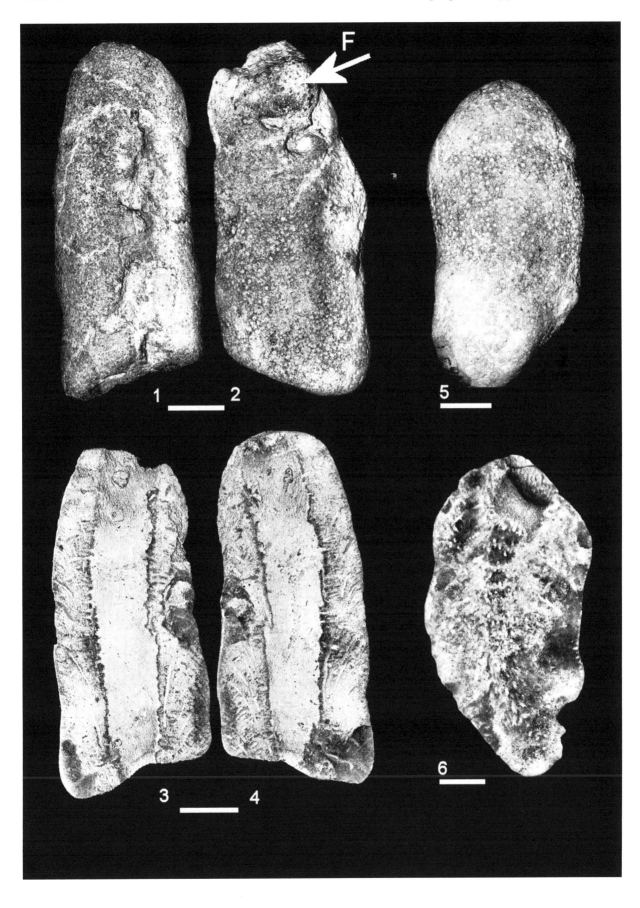

composed of a net of vertical, horizontally radiating and sub-horizontal, concentric tracts. These form sub-quadrangular openings, both horizontally and vertically (Rigby & Dixon 1979). The species is represented by one specimen, NRM Sp10156, which is a near-complete, sub-cylindrical sponge, 68 mm tall, with diameter decreasing from 30 mm near the base to 22 mm at the top (Pl. 15, figs 1–4). The sponge body has been laterally compressed and, despite abrasion, appears to be weakly annulate. On the lateral surface near the top, a hemispherical tabulate (*Favosites* sp.) has grown (Pl. 15, fig. 2, arrow F). The sponge body has been cut axially, revealing the deep spongocoel, which extends throughout the sponge body, widening from 9 mm in the lower part to 12 mm in diameter at the top. It is filled by matrix, thus obscuring the gastral surface (Pl. 15, figs 3–4), although it is visible in cross-section. The lateral surface is covered by randomly scattered ostia, 0.3–0.5 mm in diameter. Skeletal tracts are better preserved dermally than internally.

There are three canal systems, each of which is connected with the others: vertical; sub-horizontal radial, and horizontal-concentric. Vertical canals run initially parallel to the spongocoel, then rise upward and outward to meet the dermal surface at low angles, between 15 and 20°. The larger canals, 0.5 mm in diameter, are separated from each other by smaller, parallel canals approximately 0.3 mm in diameter.

The second canal system consists of parallel, sub-horizontal radial canals, on average 0.3–0.5 mm across, perpendicular to the vertical canals. They extend upwards from the dermal zone, to a maximum height at around mid-wall, and then run horizontally or sometimes curve slightly downwards, where they empty into the spongocoel. Most of these canals empty normal into the gastral wall, but some meet the gastral wall at an angle of about 80°. The gastral wall is not smooth, but shows irregular projections where radial canals empty into the spongocoel.

Horizontal-concentric canals are smaller and less clearly preserved. They appear in cross-section as elliptical, 0.15–0.30 mm (averaging 0.20 mm) in diameter. A precise, detailed description is not possible, but they seem to be in open connection with the sub-horizontal radial canals.

The dominant skeletal structures are tracts formed from rhizoclones as described above; both vertically and horizontally arranged tracts are present.

According to Rigby & Dixon (1979), tract diameter increases from the interior towards the dermal wall. In the present specimen, tracts and mesh spaces could be measured only at the dermal surface, where a small region is preserved in translucent chalcedony (Pl. 15, fig. 4, lower right). Horizontal and vertical tracts are of approximately equal dimensions, varying from 0.05 to 0.20 mm, but usually 0.10 mm across. Sub-circular mesh spaces are 0.08–0.15 mm in diameter, and elliptical mesh spaces are 0.15–0.25 mm across. Some sub-quadrangular meshes occur, with sides 0.10 mm long. Individual rhizoclones could not be observed.

Remarks. – According to the classification used by Finks (1960, p. 88, see above), *H. cylindricum* falls into his group 1A, together with *H. minutum* Rigby & Dixon, 1979, *H. armstrongi* Young & Young, 1877, *H. aeluroglossa* Finks, 1960 and *H. toftanum* sp. nov.

In *H. armstrongi*, tracts are composed of numerous spicules. *H. minutum* is a massive sponge, lacking a well developed canal system. *H. aeluroglossa* is a flattened spheroidal to lobate or digitate sponge, without a spongocoel. *H. cylindricum* differs from these species in its cylindrical body shape and the deep, wide spongocoel extending to the base. *H. toftanum* sp. nov. is also cylindrical, but its canal system is different, consisting of a bundle of axial canals, with large, convex or S-shaped convergent canals and smaller, convex, diverging canals.

Haplistion toftanum n. sp.

Plate 15, figures 5–6; Plate 16, figures 1–7

Derivation of name. – Named after Tofta parish on Gotland, Sweden, from where numerous sponges in the present assemblage have been collected.

Diagnosis. – Obconical to (sub-) cylindrical sponge with weakly annulated exterior, on which numerous ostia are distributed randomly. Bundle of parallel axial

---

Explanation of Plate 16.
Figs 1–8. *Haplistion toftanum* n. sp.
1–2, 8, NRM Sp10173; paratype; **1**, lateral view on dermal surface; **2**, view on axial surface; **8**, detail of base seen in Figure 2, showing diverging tracts composed of rhizoclones. **4–6**, NRM Sp10181; paratype; **4**, view of upper surface with sinuous, radiating canals running to the axial zone; **5**, lateral view of dermal surface, with large parietal gaps (see text); **6**, numerous radial canals in cross-section or oblique section. Canals in the left part are filled with white chalcedony. **3, 7**, NRM Sp10185; paratype; **3**, polished axial plane. Convex canals converge to the axial region, flaring upwards in the axial region, as shown at upper left; **7**, counter-part of **3**; magnified, oblique cross-section of canals in the axial region. Walls of canals are pyritized.
Scale bars represent 1 cm (Figs **1–6, 8**) and 1 mm (Fig. **7**).

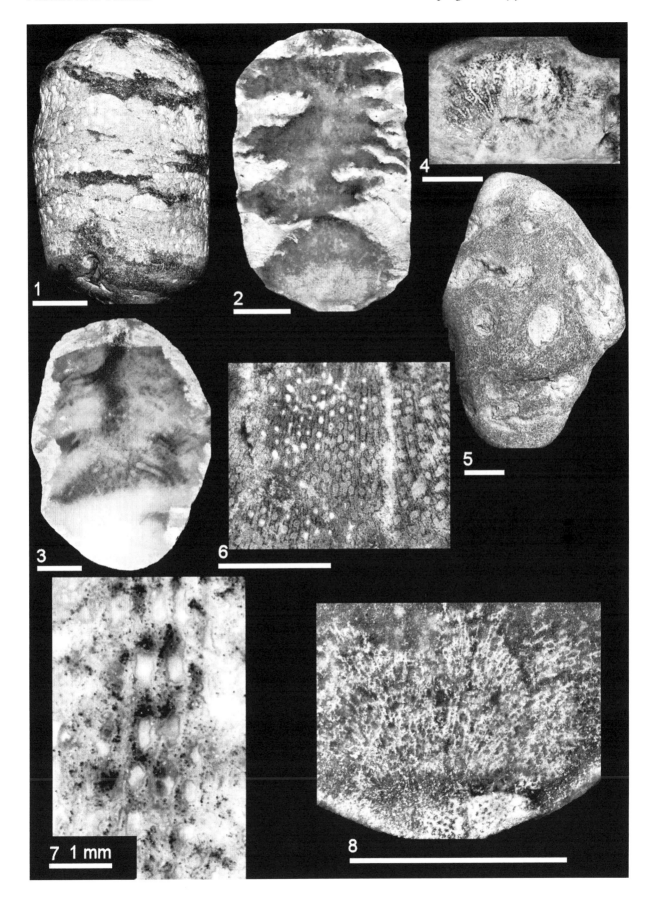

canals running from the base to the top, emptying into a shallow spongocoel, or spongocoel absent. Large, convexly curved canals run from the exterior upward and inward to the axial region, emptying into an axial canal; or flaring upwards, concavely curved from about mid-wall, to continue as axial canal. In a second system, convexly curved, parallel canals diverge from the axial region to the distal region, where they empty at high angles or normal to the lateral surface. Skeletal net of tracts of equal dimensions, composed of complexly sculptured rhizoclones; sub-rounded mesh spaces, their diameter increasing towards the exterior. Monaxons and dendroclones probably absent.

Holotype. – NRM Sp10157 (Pl. 15, figs 5–6) is an erratic from pebble accumulations on the coast at Blåhäll/Tofta (Gotland, Sweden), coordinates: 57°31′ N, 18°06′ E; specimen cut axially.

Paratypes. – NRM Sp10173, NRM Sp10181 and NRM Sp10185; collected from Blåhäll/Tofta and Sp10293; from Gnisvärd. All have been cut axially.

Other material. – NRM Sp10291; from Stenkyrkehuk. NRM Sp10292; from Gnisvärd. NRM Sp10294 is from Blåhäll/Tofta. All have been cut axially.

Description. – NRM Sp10157, the holotype, is an irregularly sub-cylindrical sponge, 37 mm long, 28 mm wide and 70 mm high (Pl. 15, figs 5–6). The exterior shows alternating smooth horizontal crests and depressions, separated by approximately 25 mm between crests, the depth of which have been reduced to 2 mm due to abrasion. The dermal surface (Pl. 15, fig. 5) shows randomly scattered possible apopores, appearing as pits 0.3–0.5 mm wide, most of them 0.4 mm in diameter. Between them are smaller pores, possibly prosopores, 0.25–0.30 mm in diameter; these appear as brown spots of chalcedony and are about twice to three times as numerous as the pits. The annulate character is also recognizable in NRM Sp10173, despite being severely eroded (Pl. 16, fig. 1). It is a complete, although worn, cylindrical specimen, 38 mm long, 28 mm wide and 60 mm high, and has no spongocoel. The transition from the convexly rounded base, with concentric annuli, to the lateral surface is rather abrupt. The transition from the side to the rounded top is more gradual (Pl. 16, fig. 2). The axial plane reflects alternating zones visible on the surface, in different colours and density of chalcedony. This specimen appears to be horizontally layered, showing irregular, alternating light-grey and bluish-grey zones, and externally shows four horizontal, regular sets of crests and depressions, about 11 mm between crests. In this respect, it is almost

identical to NRM Sp10185 (see below). The surface of NRM Sp10181 differs from the other specimens in having large hollows distributed over the surface, 7–14 mm in diameter, and interpreted as parietal gaps (Pl. 16, fig. 5), but no relationship to the internal canal system could be established. The hollows are filled with matrix. The specimen shows no spongocoel; the top shows sinuous, radial canals, apparently converging to and emptying into large vertical, axial canals (Pl. 16, fig. 4).

A deep, central spongocoel is absent in *H. toftanum* n. sp., although a shallow one, 11 mm wide and 12 mm deep, is present in the holotype (Pl. 16, fig. 2). In NRM Sp10173 and NRM Sp10185, a bundle of axial canals open at the rounded apex (Pl. 16, figs 2–3).

Three canal systems combine to form the aquiferous system. Axial canals, 0.6–0.8 mm in diameter, run from the base to the top. Large, convergent canals, 0.6–0.9 mm across, rise both from distal parts and from deeper internal areas; these run upwards to the axial zone, bending convexly. In the lower part of the sponge body, they probably empty into an axial canal. In the upper part of the sponge body, these canals are S-shaped, being initially bent convexly, but near the axial zone they curve upwards concavely, to continue as vertical, axial canals (Pl. 15, fig. 6; Pl. 16, fig. 3). The initiation of some of these canals is unusual: they start very small, but within a distance of 1 mm, they increase in diameter abruptly from 0.01 mm to 0.8 mm. The canals are clearly visible preserved as translucent chalcedony in NRM Sp10157. These canals are separated vertically from each other by three to six smaller, parallel canals, about 0.15 mm in diameter, which also empty into the axial canals. Cross-sections of these appear distally as spots of translucent chalcedony (Pl. 16, figs 6–7).

Perpendicular to the convex, converging canals is a system of parallel, diverging canals, rising from the base and the axial zone, and curving convexly upwards and outwards to open into the lateral surface. These canals are relatively large, at 0.4–0.6 mm in diameter. In the lower part, they empty at right angles to the dermal surface, and in the upper part at angles between 70 and 90°, where their corresponding ostia form small pits (see above).

The skeleton is formed by tracts composed of subparallel rhizoclones. Tracts are arranged three-dimensionally, forming canal walls and enclosing a network of mesh spaces. Tract diameters vary from 0.03 to 0.05 mm and are equal in all directions. Mesh spaces could be measured in the interior of NRM Sp10157 and in the basal part of NRM Sp10173 (Pl. 16, fig. 8). They are sub-circular (on average 0.10 mm in diameter), elliptical (measuring 0.10–0.15 mm) or sub-quadrangular with rounded corners (0.15 mm). Mesh spaces in the distal zone could not be measured precisely, but are somewhat, though not substantially, wider than those in the interior.

Isolated rhizoclones could only be observed occasionally. Where visible, they are 0.20–0.35 mm long, with an axis 0.025–0.040 mm in diameter, and elaborately sculptured zygomes. Some bridging dendroclones, 0.15–0.40 mm long and 0.03–0.05 mm in diameter, occur in the basal regions of NRM Sp10173 and NRM Sp10185. Monaxons have not been observed.

Remarks. – Most of the internal structures have been destroyed by intensive silicification and replacement by chalcedony. In addition, all specimens have been intensely abraded, so it is possible that a distinct dermal layer has been lost. Nevertheless, several distinctive characteristics justify the erection of a new species. The diameter of tracts being of equal dimensions, and the mesh spaces being smaller than 0.50 mm, dictates that *H. toftanum* sp. nov. falls into Finks' group 1A (Finks 1960, see above).

According to the brief description by Young & Young (1877, p. 428), *H. armstrongi* is a small, subcircular sponge, specimens of which are about 15 mm high and 10 mm wide, without the complex canal system as it is developed in *H. toftanum* n. sp.

Haplistion minutum Rigby & Dixon, 1979 is distinguished by its massive body shape, its rectangular mesh spaces, the occurrence of bridging dendroclones and the absence of a conspicuous canal system.

The body shape of the Devonian *H. aeluroglossa* Finks, 1960 is flattened ovoid to lobate or subdigitate and the species has larger rhizoclones (0.3–0.8 mm long, 0.05 mm in diameter). It also differs by the occurrence of oxeas in tracts, and rectangular radial canals.

Haplistion frustrum Rigby & Chatterton, 1989 is discoid to funnel-shaped, with all tracts of about the same diameter, and is therefore assignable to group 1, but with large mesh spaces, 0.5–0.8 mm, assignable to group 2A. In addition, cross-bracing dendroclones are common in *H. frustrum*, but are rare in *H. toftanum* n. sp.

The dimensions of rhizoclones and the tract diameters in *H. regularis* Rigby & Webby, 1988 are similar to those in *H. toftanum* n. sp. However, in *H. regularis,* the dominant radial tracts are twice the diameter of horizontal tracts, which is a characteristic of Finks' group 2. In addition

H. regularis commonly has cross-connected dendroclones.

Haplistion cylindricum Rigby & Dixon, 1979 is the most similar species in terms of body shape and skeletal structure, but it has a long, deep, open spongocoel, extending to the base of the sponge body, as shown in NRM Sp10156 (see above, Pl. 15, figs 3–4). In addition, its system of large, long, diverging canals, flaring upwards and outwards, meeting the dermal wall at low angles, is distinct from that in *H. toftanum* n. sp.

Haplistion sp. Young & Young, 1877

Plate 17, figures 1–3

Material. – NRM Sp10183 is an abraded, sub-cylindrical sponge from pebble accumulations near Blåhäll/Tofta.

Description. – The sponge body is irregularly curved, and measures 67 mm in height and 30 × 35 mm in diameter. It has been cut twice, longitudinally through the axis, and transversely near the top. One side of the lateral surface is characterized by large, sub-parallel but irregularly winding vertical canals, 1.5–3.0 mm (on average 2 mm) in diameter (Pl. 17, fig. 1). Canals on the opposite surface are of the same dimensions, but are considerably less regular and locally anastomosing. Approximately eight canals meet the rounded upper surface, but without showing any convergence or indication of a spongocoel (Pl. 17, fig. 3). There is no difference in directionality, dimensions or arrangement regularity between these canals and those on the lateral surface.

The skeleton is preserved very poorly, with only locally recognizable spicule tracts and canals. The skeleton is a meshwork of tracts, each composed of tiny rhizoclones. Mesh spaces enclose small radial canals, 0.1–0.2 mm in diameter, distally 0.2–0.6 in diameter. They are arranged pinnately, initially flaring upward and outward at ca. 30°. Towards the distal zone, they curve more strongly outward, to meet the surface at right angles (Pl. 17, fig. 2). Rare dendroclones occur locally.

The diameter of tracts is approximately uniform throughout the sponge body, varying from 0.03 to 0.05 mm. Isolated rhizoclones could not been measured. Isolated rhizoclones are locally visible on the lateral surface as remains of tracts, which are pyritized and for the greater part dissolved. Even where present, however, they are too deformed diagenetically to measure them reliably.

Remarks. – The skeletal construction of tracts that define sub-circular mesh spaces refers the specimen to *Haplistion* Young & Young, 1877. It falls within Finks' (1960) Group 1, as it possesses mesh spaces smaller than 1 mm in diameter.

Haplistion minutum Rigby & Dixon, 1979 differs from NRM Sp10183 in body form, being a massive to massive lobate sponge. *H. cylindricum* Rigby & Dixon, 1979 is cylindrical, but differs in its deep, wide spongocoel and marked annuli. The cylindrical *H. toftanum* n. sp. has a conspicuous canal system of axial, converging and diverging canals, and therefore differs considerably in internal structure. The differences with other species in *Haplistion* are even more conspicuous. Thus, NRM Sp10183 is not assignable to a known species. The specimen is left in open nomenclature, however, since although it is the only specimen with this structure so far recognized, preservation is too poor to allow a full description.

Genus *Warrigalia* Rigby & Webby, 1988

Type species. – *Warrigalia elliptica* Rigby & Webby, 1988.

Remarks. – *Warrigalia* was erected to include Late Ordovician rhizomorine sponges from the Malongulli Formation in New South Wales, Australia. It is separated from *Haplistion* Young & Young, 1877 by the latter having a more-or-less rectangular pattern of tracts, usually cored by monaxons, and forming the walls of radial and sub-horizontal canals. The same is true of *Chaunactis* Finks, 1960, in which the rectangular pattern is even more distinct than in *Haplistion*, and which also has a dermal layer consisting of a rectangular network of bundles of parallel monaxons. *Kazania* Stuckenberg, 1895 differs from *Warrigalia* in having long coring monaxons in the tracts. *Parodospongia* Rigby & Chatterton, 1989 is thin-walled, cup- or bowl-shaped and has a deep spongocoel. *Columellaespongia* Pickett, 1969 is a thick-walled cylindrical sponge with a deep spongocoel. *Oremo* Pickett, 1969 is irregular to globose in body form, but the published diagnosis and description are too brief to allow a detailed comparison.

Crawneya Pickett, 1969 and *Warrigalia* are both massive, possibly initially encrusting sponges, with anastomosing tracts and canals; however, *Crawneya* is lobate, with large, dome-shaped outgrowths and reaches impressive dimensions, up to 50 cm (Pickett 1969, p. 13). Tracts in *Crawneya* are 0.5–0.7 mm in diameter, considerably thicker than those in *Warrigalia*. *Taplowia* Rigby & Webby, 1988 is a massive, obconical sponge with a dense dermal layer, and internally with horizontal, laminate elements and vertical pillars combining to form chambers (Rigby & Webby, 1988, p. 21). Thus, it is distinctly different from the massive, flabellate body form and open structure of *Warrigalia*. According to the diagnosis given by Rigby & Webby (1988, p. 22), *Lewinia* is an obconical to open disclike cavernous sponge with a skeleton dominantly of vertical rods and weblike blades, and coring monaxons in its tracts. *Boonderooia* Rigby & Webby, 1988 shows some similarities with *Warrigalia*, such as the thick-walled, flabellate body shape, but is distinguished by the occurrence of coring oxeas, which project through tracts and emerge as spines at tract junctions.

The occurrence of *Warrigalia* in the present assemblage constitutes a substantial range extension, both temporally (from Late Ordovician to Early Silurian) and geographically (first record outside Australia).

Warrigalia robusta Rigby & Webby, 1988

Plate 18, figures 1–6

1888 *Warrigalia robusta* Rigby & Webby, pp. 20–21, pl. 2, figs 5–6, pl. 3, figs 1–4.

Material. – NRM Sp10089, NRM Sp10153, NRM Sp10174 and NRM Sp10187; collected from gravel accumulations at the coast near Blåhall/Tofta.

Description. – *Warrigalia robusta* is a massive to sheet-like sponge, characterized by sub-cylindrical

---

Explanation of Plate 17.
Figs 1–3. *Haplistion* sp. Young & Young, 1877
1–3, NRM Sp10183; **1**, lateral surface with sub-parallel canals. The outgrowth in the middle part is *Favosites* sp.; **2**, longitudinal section, with large, irregularly oriented canals. Skeleton and smaller, pinnately arranged canals are poorly preserved; **3**, top view with cross-sections of large canals. Small white spots at the upper right and in the upper part in **1** are eroded tracts. Isolated rhizoclones are not recognizable.
Fig. 4. Rhizomorina indet.
4, NRM Sp10184, oblique view of presumed upper side, showing conspicuous concentric growth stages. Radiall-arranged tracts and small canals are visible at right. At left, there is a bundle of large, parallel canals in natural oblique section.
All scale bars represent 1 cm.

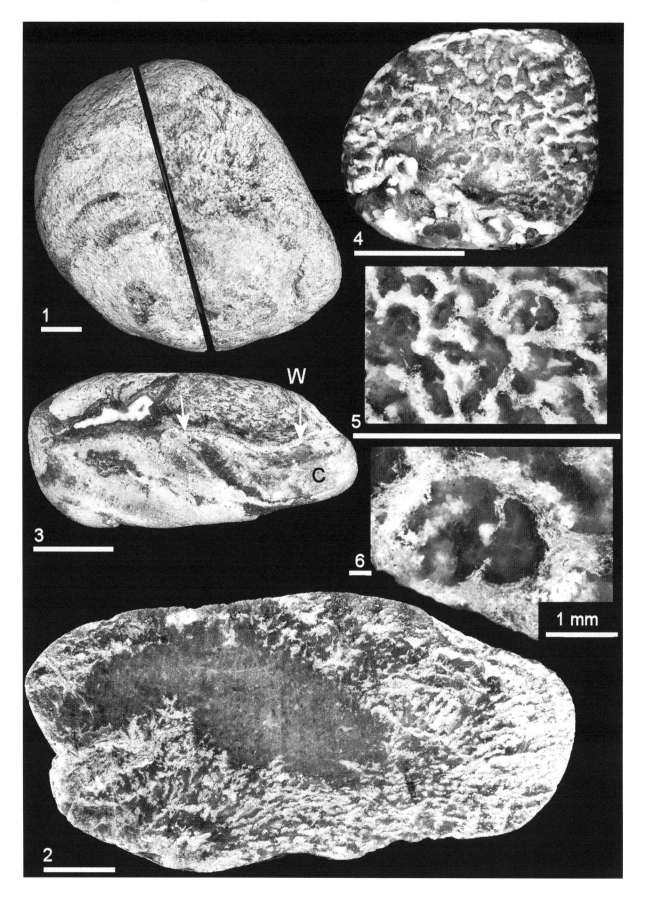

to ribbon-like vertical tracts, cross-connected by anastomosing horizontal tracts (Rigby & Webby 1988, p. 20). Four specimens in this collection have been recognized as *W. robusta*. The largest specimen, NRM Sp10089, is a flabellate, massive sponge, 78 × 62 mm in diameter and 36 mm high, and severely abraded. Irregular tracts and canals diverge in all directions from the broadly homogeneous basal region, cutting across a number of sub-horizontal, concentric structures (Pl. 18, figs 1–2). The concentric layering is accentuated by a horizontal, 12-mm-deep, 1- to 3-mm-wide slit at about half-height, filled with the same argillaceous matrix that occurs on most of the sponges. A 30 × 25 mm flake broken from the base of the specimen shows part of the internal organization. The upper surface shows no details beyond a poorly defined pattern of spicule tracts. If the sponge possessed a dermal layer, it has been destroyed by abrasion. The specimen has been cut axially, with the polished planes showing the general development and internal structure. Tracts radiate in all directions from the base, but were particularly well developed towards one side. The tracts are irregularly disposed, enclosing canals that also form a flabellate structure. Sub-horizontal banding is caused by cross-connecting tracts that are slightly curved convexly and enclose poorly defined canals (Pl. 18, fig. 2).

An encrusting growth form is demonstrated by NRM Sp10153 (Pl. 18, fig. 3, top right). This small, partial sponge body measures 21 × 12 × 8 mm and occurs in a pebble fragment measuring 34 × 14 × 5 mm. The sponge nucleated on the stromatoporoid *Clathrodictyon* sp. On one side, the sponge has overgrown a brachiopod, most probably *Pentameroides* sp. Although only a fragment, it shows the characteristic anastomosing arrangement of tracts and canals, as in the previous specimen. NRM Sp10174 has been abraded into an ovoid pebble, 51 × 30 × 24 mm, partly consisting of homogeneous chalcedony, but with some parts of the surface showing tracts with the same organization and dimensions as in the previous specimens.

Radially diverging tracts composed of numerous rhizoclones have been very well preserved in the small, strongly worn specimen NRM Sp10187

(Pl. 18, fig. 4). The sponge nucleated on the tabulate *Subalveolites* sp. Tracts are 0.25–0.50 mm wide, winding and bifurcating irregularly, and enclosing net spaces that are 0.7–1.0 mm wide and 1.5–2 mm high. They frequently merge and anastomose throughout the sponge body. Tracts may fuse lengthwise to form ribbon- or blade-like constructions (Pl. 18, fig. 5), as well as broader cyst-like plates (as described by Rigby & Webby 1988, p. 20, pl. 2, figs 5–6). Ribbons and cyst-like plates could not be measured precisely due to being obliquely embedded in chalcedony. They seem to have functioned as support or consolidation for the endosomal skeleton, whereas smaller tracts formed canal walls. A few tracts are visibly composed of (sub-) parallel rhizoclones. Horizontal tracts are poorly preserved as powdery silicified material, occasionally showing some remnant of the skeletal structure. Sub-vertical, elliptical to circular canals run subparallel to the general direction of growth and are 0.5–1.4 mm in diameter. They do not reach the surface, probably ending in the dermal zone, but their precise terminations could not be established. Sub-circular canals run normal to the sub-vertical canals and are 0.5–0.8 mm in diameter.

Individual rhizoclones are 0.13–0.33 mm long (mean 0.22 mm) and 0.015 mm wide (Pl. 18, fig. 6). They have elaborate extensions on both sides of the axis, if attached to adjacent spicules to form part of a tract. Rhizoclones that occur as part of a canal wall have a smooth concave side and extensions on the convex side. The heloclone-like spicules described by Rigby & Webby (1988) have not been found.

Remarks. – Most of the skeleton in the present specimens has been dissolved and replaced by chalcedony, in contrast with the type material from New South Wales, in which the tracts are free of matrix. Isolated rhizoclones can be recognized only locally, and the same is true of rhizoclones arranged in tracts. As a result, details of the composition of the ribbon-like and cyst-like constructions remain largely obscure. Nevertheless, no other Palaeozoic rhizoclonid sponges yet known share the characteristic features of *W. robusta* seen in the specimens discussed here.

Explanation of Plate 18.
Figs 1–6. *Warrigalia robusta* Rigby & Webby, 1988
1–2, NRM Sp10089; **1**, oblique view of base and lateral surface; **2**, longitudinal section showing the irregular direction of growth. Tracts are irregularly disposed and anastomose, and do not produce a network of canals as in *Haplistion*; **3**, NRM Sp10153. A small fragment of the sponge body ('W') at the top. Arrows indicate the connection zone with *Clathrodictyon* ('C'). At upper left, a small brachiopod has been overgrown by *W. Robusta*; **4–6**, NRM Sp10187; **4**, View of polished surface, showing the robust, irregular, thick tracts; **5**, Detail of Figure **4**, showing the irregularity of tracts; **6**, detail of Figure **5**, showing the fibrous character of the tracts, as a stage in the diagenetic process of partial tract dissolution and replacement by chalcedony. In the eroded margins, small numbers of connected (and some isolated) rhizoclones are visible.
Scale bars represent 1 cm (Figs **1–5**) and 1 mm (Fig. **6**).

The supposed lack of monaxons in *Warrigalia*, as stated in the generic diagnosis, is contradicted by examples in Rigby & Webby (1988, p. 19, text–fig. 6B). The rare occurrence of monaxons in the specimens discussed here therefore poses no problem for the assignment to *W. robusta*, and the current material is not separable from the Australian specimens.

Warrigalia elliptica Rigby & Webby 1988 differs from *W. robusta* in its more delicate spicule tracts and more regularly disposed canals, which are consistently elliptical.

Rhizomorina indet.

Plate 17, figure 4

Material. – NRM Sp10184, collected from Blåhäll/Tofta, is a unique specimen that differs from the Rhizomorina described above. Part of the sponge has been sectioned vertically.

Description. – The specimen is a fragment of a probably discoidal sponge, originally about 110 mm in diameter, 25 mm tall at the centre and 17 mm at the edge. The flatter surface, presumed to be the base, shows a somewhat radial pattern of tracts, which is often interrupted by small canals and the tracts distributed apparently randomly. The opposite side is also more-or-less flat and shows distally an irregularly concentric, wrinkled pattern of growth lines. Perpendicular to the flat surfaces is an irregularly radial pattern of canals and tracts (Pl. 17, fig. 4). The central part differs conspicuously in form from the flatter distal zone, arching gradually upwards until reaching a thickness of about 25 mm. This presumably represents the original centre of the sponge body and contains 0.50–1.0 mm-wide sub-parallel canals, arranged obliquely. Slightly sinuous skeletal tracts separate them from each other, 1–4 mm wide. The large canals are not radially arranged nor do they arch up acutely. Considering the general level of abrasion of sponges in this assemblage, the sponge body was probably rather more discus-shaped than dome-shaped or hemispherical.

All visible skeletal structures are restricted to the surface, as chalcedony replacement has destroyed all useful internal details. Remains of the skeleton have occasionally been preserved in the transitional zone from the eroded surface into the internal dense chalcedony, which is best exposed in some parts of the natural cross-section. Tiny rhizoclones are the only type of spicule that could be observed. Tracts are coarse, thick and very irregularly arranged. Tracts in the distal zone enclose numerous small, radially arranged canals, 0.25–0.30 in diameter. Tracts in the central part are considerably thicker, enclosing much larger canals, 0.50–1.0 mm across, that are generally arranged sub-parallel to the base.

Remarks. – Numerous washed-in monaxons and hexactines are visible on the supposed upper surface, confirming that it is part of the Telychian assemblage.

Skeletal characteristics assign the sponge body to the suborder Rhizomorina Zittel, 1895. Other rhizomorine genera in the present assemblage are *Haplistion* Young & Young, 1877 and *Warrigalia* Rigby & Webby, 1988. NRM Sp10184 differs conspicuously from *Haplistion*, in which tracts are considerably thinner (Finks' Group 1), and the general body shape varies from cylindrical to hemispherical.

NRM Sp10184 seems to be more closely related to *W. robusta* Rigby & Webby, 1988. It shares the coarse, very irregular spicule tracts, but differs too much in the arrangement of canals, tracts and body form to assign it to *Warrigalia*. The specimen possibly represents a new taxon, but a more detailed systematic treatment would be overly speculative, considering the poor preservation and rarity of material.

There are probably additional rhizomorine specimens in the collections that have as yet remained unrecognized. They may also occur among the many hundreds of specimens of the Ordovician *Aulocopium aurantium* Oswald, 1847, or among the sponges that currently remain unidentified.

Subclass Tetractinellida? Marshall, 1876

Discussion. – Although traditionally placed in the subclass Tetractinomorpha, recent molecular phylogenetic work (Borchiellini *et al.* 2004) has redefined demosponge evolutionary topology. In their phylogenetic terminology, the Hadromerida are included in Borchiellini *et al.*'s Group G4, However, Erpenbeck & Wörheide (2008) found that tetractine-bearing lineages were highly restricted and form a discrete clade, suggesting that they may have evolved after separation of the other lineages. As the clade definition of Tetractinellida given by Borchiellini *et al.* (2004) was based on the first appearance of spicules synapomorphic with those of geodiid tetractines, it is not possible to recognize this grouping until we know the point of first appearance of tetractines. For the moment, we tentatively assign the Hadromerida to the Tetractinell-ida, but recognize that the current definitions are potentially unworkable.

Order Hadromerida? Topsent, 1898
Family uncertain

Remarks. – Abundant specimens of this monaxon-based sponge are found attached primarily to lithis-tid sponge skeletons. Some include a few large hexactine-based spicules within the structure, but these are likely to be detrital. All share a similar architecture, with spicules irregularly radiating from an attachment surface, to some extent in tufts, but merging to form a dense coating over the affected area of substrate.

The fundamental architecture of this sponge is similar to that of a number of taxa described from the Devonian and later periods and included in the Sollasellidae (Finks & Rigby 2004a). However, the Sollasellidae is an extant monogeneric family of complex internal organization (van Soest 2002). Although they include radial fabrics in part, the fossil sponges more closely resemble other extant families in which the structure is a simpler, purely radiating array of monaxons, such as the Tethyidae or Hemiasterellidae (Hooper 2002; Sarà 2002). These hadromerid families are also characterized by asterose (often hexactinal) micro-scleres, which might potentially be evolutionarily linked to hexasters in the stem-group hexactinellids. It is, however, unclear whether there is a phylogenetic link between the extant taxa and the fossil groups, as the radiate architecture could be convergent and there is yet no evidence of micro-scleres in the fossils. Assignment to particular extant families is unwarranted from the level of detail normally available, even if a close phylogenetic link is established.

Genus *Opetionella* von Zittel, 1878

Emended diagnosis. – Globular tuberous to irregularly crustose appearing; neither oscula, pores, nor canals observed; skeleton a thick layer of closely spaced, parallel oxeas or styles (emended from Finks & Rigby 2004a, p. 21).

Remarks. – Several structurally similar taxa were included by Finks & Rigby (2004a) in the Sollaselli-dae (discussed above). Most of these taxa (e.g. *Coniculospongia* Rigby & Clement, 1995, *Ginkgospongia* Rigby & Clement, 1995 and *Sphaeriella* Rigby & Pollard-Bryant, 1979) differ fundamentally in the body

form, showing a conical, spherical or stalked growth form. The related Tethyidae are also represented by dominantly spherical fossils, with their radial monaxons strongly bundled, but they are only known from the Cenozoic (Finks & Rigby 2004a). Of the remainder, two are potentially similar to the current sponges: *Opetionella* Zittel, 1878b and *Trichospongia* Billings, 1865. *Trichospongia* is poorly known, but appears to have been hemispherical, and presumably encrusting. However, Finks & Rigby (2004a) include clear branching canals in the diagnosis, together with a weak concentric structure, neither of which are seen in the irregular morphology of the present material. The Jurassic *Opetionella*, however, is a crustose to globular, tuberose sponge that lacked any obvious osculum, pores or canals. There are differences in specific features of the skeleton, but these are very minor, and despite the remarkable age difference between the occurrences, the two taxa must, at least presently, be included in the same genus.

Opetionella incompta n. sp.

Plate 19, figures 1–8

Derivation of name. – From *incompta* (Latin), meaning unkempt or untidy; referring to the appearance of the dense encrusting growths that irregularly cover lithistid surfaces and other substrates.

Holotype. – NRM Sp10274 (Pl. 19, figs 1, 2, 5), a typical specimen preserved in the external grooves of a specimen of *Caryoconus gothlandicus*, and showing extensive areas with clearly preserved spicules and the translucent chalcedony structures at the base that are interpreted as representing the sponge soft tissue body. Transported erratic specimen from gravel accumulations at the coast at Blåhäll/Tofta (Gotland, Sweden), coordinates 57°31′ N, 18°06′ E.

Paratypes. – NRM Sp10083, NRM Sp10275, NRM Sp10281, NRM Sp10282, NRM Sp10283, NRM Sp10285, NRM Sp10289 and NRM Sp10290; a range of examples illustrating the variation present in the material.

Other material. – NRM Sp10065, NRM Sp10287 and NRM Sp10288: probable specimens with somewhat unusual features, such that their inclusion is tentative. Partial specimens are also present on many other sponges in the collections.

Diagnosis. – *Opetionella* with spicules relatively slender styles, in large specimens at least 4 mm long

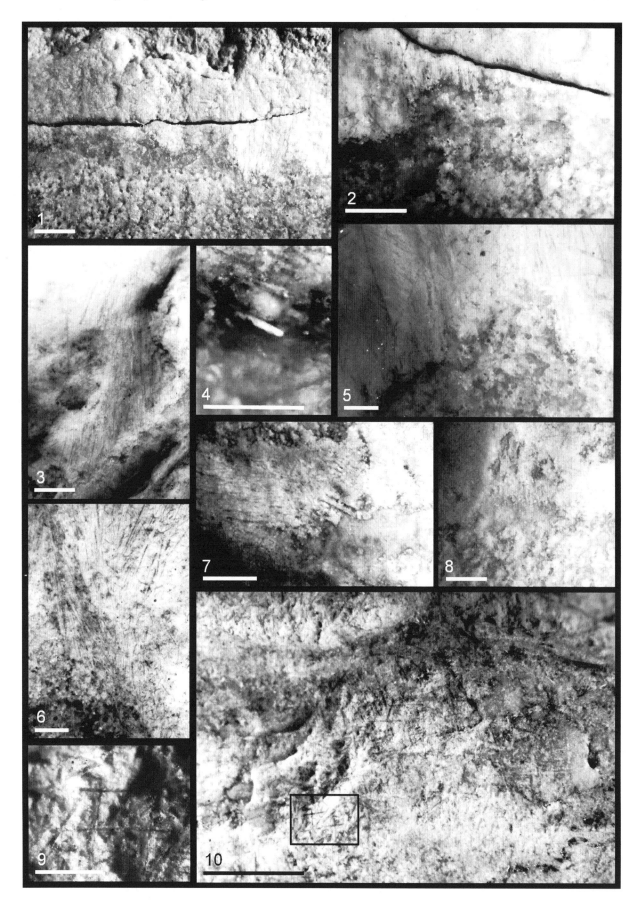

(perhaps over 10 mm long) and with maximum diameter up to 0.08 mm; sponge thickness irregular and soft-tissue layer probably very thin (approximately 1 mm); in mature specimens, monaxons extend beyond soft tissue to form long, hair-like array.

Description. – *Opetionella* is primarily found as mats encrusting the surfaces of a variety of lithistids, particularly in surface grooves or cavities such as the osculum. In rare specimens (Pl. 19, fig. 3), the sponge grew over a brachiopod fragment or other hard surfaces. Coverage is irregular, and in one specimen (NRM Sp10289, Pl. 19, fig. 6), the sponge coats one side of the osculum, but not the other and also appears to be overgrowing a thin surface-parallel mat of monaxons representing detritus from a previous phase of encrustation. Poorly preserved specimens and masses of disarticulated monaxial spicules are abundant in the samples, suggesting that the species was near-ubiquitous within the assemblage.

Several specimens are seen along grooves in the surface of various lithistids (particularly *Caryoconus gothlandicus*), with spicules generally preserved flattened onto the surface, and perpendicular to the groove, growing in from both sides. Monaxons are of variable size, with the larger specimens having spicules in excess of 3 mm long (often >4 mm long, and in NRM Sp10285, 6 mm) and 0.05–0.08 mm in maximum diameter. Entire spicules are not seen due to the three-dimensional preservation with surfaces broken or abraded through the architecture. A few spicule bases are preserved in clear chalcedony in NRM Sp10290, where their terminations are seen to be rounded. Occasionally, distal ends of spicules are visible and are always sharply pointed; the spicules are therefore styles.

The only spicules seen to originate above the nucleation surface are aligned at irregular angles and may be displaced or remobilized fragments from disarticulated skeletons. This allows estimates of the absolute lengths of spicules, as coherent swathes of monaxons are more easily measured than individual spicules; this leads to maximum length estimates of 8 mm in many areas of unconstrained growth; in

some specimens (NRM Sp10283, NRM Sp10281; Pl. 19, fig. 7), approaching 15 mm. Smaller spicules show no clear size ordering, and the smallest spicules are around 0.01 mm in diameter and less than 1 mm long.

In some cases (e.g. NRM Sp10281, NRM Sp10275; Pl. 19, fig. 8), the basal parts of the sponge are preserved with cuspate areas of clear chalcedony, approximately 1 mm tall, attached directly to the surface and presumably representing soft tissue. This, combined with the apparently disturbed or swept-back, non-orthogonal arrays of spicules seen in many specimens (e.g. NRM Sp10289), suggests that the soft-tissue regions were limited, and the spicules projected from the surface as a dense protective fur. This also explains the lack of spicules originating higher in the array, with irregular spicules in the distal parts of the skeletons almost certain to be foreign or detached fragments randomly tangled into the projecting array.

Remarks. – Differences from other genera are described under the generic remarks. It is differentiated from the type species, *O. radians* von Zittel, 1878b, by the narrower spicule proportions, and by the fact that the spicules are styles rather than oxeas.

The sponge appears to have been extraordinarily abundant, to the extent of perhaps binding the surface of loose fragments together and therefore potentially making a significant impact on the nature of the substrate. The density of disarticulated spicules in the sediment would also have affected the sediment consistency, perhaps encouraging attached sessile suspension feeders over mobile endobenthos.

Class Hexactinellida Schmidt, 1870
Order Lyssacinosida? von Zittel 1877
Family uncertain

Genus *Urphaenomenospongia* n. gen.

Derivation of name. – After Goethe's *Urphänomen*, which describes a 'part that contains the whole'

Explanation of Plate 19.
Figs 1–8. *Opetionella incompta* n. sp.
1, 2, 5, NRM Sp10274; holotype. Areas of the extensive holotype specimen showing localized development of discrete expansions of the body, and parallel array of monaxons. **3**, NRM Sp10285, showing attachment to brachiopod shell fragment. **4**, NRM Sp10290, showing rounded basal termination of styles, embedded in clear chalcedony. **6**, NRM Sp10289, with spicules swept back from encrusted oscular surface, and many fragments of broken spicules in distal areas. **7**, NRM Sp10281, showing long monaxons arising from cuspate encrusting body. **8**, NRM Sp10275, showing spicule tuft with basal soft tissue area.
Figs 9–10. Hexactinellid indet. B
9–10, NRM Sp10277; **9**, detail showing locally preserved regular arrangement of stauractines; **10**, overall view of specimen with box showing region illustrated in Figure 9.
Scale bars represent 1 mm (Figs **1–9**) and 5 mm (Fig. **10**).

(Seamon & Zajonc 1998), in reference to the detailed structural interpretation that has been possible from only a small piece of the sponge.

Diagnosis. – Thin-walled lyssacine hexactinellid with several layers: dermal, short-rayed orthpentactines overlying a sparse array of long, thin diactines; central lamina with irregularly oriented, thin-rayed hexactines and monaxons; possible gastral layer with monaxons irregularly distributed among centres of large, robust hexactines, one ray of which projects through the dermal wall to form prostalia.

Remarks. – Although only a partial specimen is available, the material preserves a complex skeletal structure that is otherwise unknown and requires recognition as a new genus. The structure is partly reminiscent of many modern hexactinellid groups, with a similar broad arrangement of spicules through the body wall seen in the euretid hexactinosan *Chonelasma* (Reiswig & Wheeler 2002). This similarity includes the combination of surface pentactines with internal hexactines and diactines, but it differs fundamentally in that the spicules in the new species are unfused. The modern Lyssacinosida are characterized by unfused spicular skeletons, often with large numbers of monaxons (e.g. Rossellidae; Tabachnik 2002). The Rossellimorpha, however, are known with certainty only since the Mesozoic (Brückner & Janussen 2005; Brückner 2006).

Although 'Lyssacinosan' sponges are sometimes regarded as being abundant in the Palaeozoic (Brückner & Janussen 2005), this refers to a descriptive definition (unfused hexactinellid-like spicules) rather than being a taxonomic grouping. The ancestral condition of hexactinellids was to have unfused spicules (see, discussions of early phylogeny: Rigby & Collins 2004; Botting & Butterfield 2005), so that lyssacine-like architecture dominates the stem group representatives. There is now known to be a molecular long branch leading to crown-group Hexactinellida (Dohrmann *et al.* 2008), implying that there was a long history of stem-group taxa. Recognition of Palaeozoic members of the living classes is difficult, largely because most shared an unfused skeleton, and micro-scleres are rarely preserved. Combinations of megasclere morphology can occasionally be diagnostic at species level, and characteristic at higher levels, but this is rarely the case even among extant groups.

The new genus includes a range of megasclere morphologies that are not specific to a particular group, and the architecture is unique and therefore also non-diagnostic. Although elements are shared with specific extant taxa, they are insufficient for a precise assignment to any group, and it may represent a late stem-group hexactinellid, or a stem-group

member of one of the subclasses. The lack of fusion of spicules implies that it is not a crown-group hexactinosan, lychniscosid or aulocalyoid. It is possible that the genus represents an early amphidiscosan, but none of the living families are closely similar in spicule complement or skeletal structure (Reiswig 2002). A general division into dermal and internal skeletal layers is typical of virtually all extant hexactinellid groups, but is unusual among Palaeozoic hexactinellids (although it is found in, for example, Brachiospongiidae; Finks & Rigby 2004b). Overall, the greatest similarity to known groups is seen with members of the modern Lyssacinosida (in particular Rossellidae and Euplectellidae), in the unfused skeleton, presence of several different megasclere morphologies, and segregation of these morphologies into different positions within the body wall. A particular feature shared with the Euplectellidae is the presence of large choanosomal hexactines with prostalial distal rays (Tabachnik 2002). Dohrmann *et al.* (2008) found the Euplectellidae to be the basal family within Lyssacinosida, but with the group as a whole relatively derived within hexactinellids. We assign the new genus tentatively to the Lyssacinosida, where it probably occupies a relatively basal position (perhaps in the stem group of Euplectellidae), but as discussed above, it is not currently possible to rule out an earlier placement within Hexactinellida.

Type species. – *Urphaenomenospongia euplectelloides* n. sp.

Urphaenomenospongia euplectelloides n. sp.

Plate 20, figures 1–4

Derivation of name. – After the structural resemblance to some living Euplectellidae.

Holotype. – NRM Sp10278 (Pl. 20, figs 1–4); partial specimen preserving detailed structure of body wall in three dimensions. Only known erratic specimen transported from gravel accumulations at the coast at Blåhäll/Tofta (Gotland, Sweden), coordinates 57°31′ N, 18°06′ E.

Description. – Portion of sponge wall, 10 mm long and 2 mm thick, containing fully articulated spicules encased in clear chalcedony. The body wall fragment is seen in section and is very slightly and sigmoidally curved (Pl. 20, fig. 1), suggesting a substantially larger overall body size. Spicules are generally preserved as outer laminae only, with extensive axial dissolution, but a few are preserved nearly intact, with relics of axial canals. The translucency of the surrounding chalcedony allows the three-dimensional structure of the skeletal body wall to be reconstructed.

The spicules are arranged in discrete layers with differing morphologies concentrated at different levels. Dermalia are largely orthopentactines (Pl. 20, fig. 2), with short, straight rays oriented parallel to surface but randomly aligned to produce an irregular, dense grid. Proximal rays (up to 1.4 mm long) roughly four times as long as lateral rays (up to 0.3 mm), straight, and projecting through to at least central spicule layer. All rays are approximately equal (up to 0.03 mm) in basal diameter. Immediately beneath the pentactines is an irregular layer of monaxons, aligned sub-parallel to the surface but in random orientation within the plane, and interwoven with the proximal pentactine rays. Monaxons are generally straight and relatively fine (typically 0.01, up to 0.02 mm in diameter); most are incomplete, but some are longer than 0.8 mm.

Central layer (Pl. 20, figs 2, 3) of skeleton composed of a loose array of monaxons (presumed diactines, but both ends of a spicule not visible) with irregularly oriented, thin-rayed hexactines (basal diameter 0.01–0.02 mm). The hexactines sometimes have four rays parallel to the surface of the sponge, but in general are irregularly oriented. Maximum observed ray length is 0.9 mm for lateral rays, but substantially shorter for distal (0.6 mm) and proximal rays (0.3 mm). Monaxons are also often somewhat inclined to the surface-parallel plane; most are incompletely preserved, but closely resemble those of the dermal layer. This layer is the least continuous and least dense, being intermittent in places (Pl. 20, fig. 3); it is generally separated from the dermal layer by between 0.3 and 0.5 mm of open space, pierced only by radial rays of spicules in other layers and occasional obliquely oriented rays from spicules in the central layer.

The presumed gastral (atrial) layer (Pl. 20, fig. 4) consists of dense, irregularly arranged but generally wall-parallel monaxons and fine-rayed hexactines, combined with the centres of large hexactines. The largest hexactines have wall-parallel rays 0.5 mm long and with basal ray diameter 0.04–0.06 mm, tapering evenly to sharp points. Distal rays (Pl. 20, fig. 2) are robust (up to 0.8 mm in basal diameter) and elongated, projecting through the remaining wall thickness and extending up to 0.6 mm as hypodermalia prostalia (total length of distal ray up to 2.0 mm). Proximal rays probably also extend through the atrial surface, but details are unclear; ray bases are robust, but are broken or obscured beyond 0.4 mm proximally. The exact position of the atrial surface is uncertain, as there is no clear spicule-bounded plane in this inner layer, although the hexactine centres coincide broadly with a lamina composed of locally densely packed monaxons. The monaxons are fine, probable diactines (as above), and include apparently curved ray fragments at variable angles to the skeletal plane. No micro-scleres seen. No spicules show spines or other notable modifications.

Remarks. – Although many Ordovician–Silurian hexactinellids have dermal pentactines, which would be the spicules most easily visible in compressed material, none described so far possess the structure described here. The Brachiospongiidae include relatively thin-walled forms with dermal differentiation of spicules, but choanosomal spicules in that group are dominantly simple hexactines. The combination of fine monaxons with robust and thin-rayed hexactines and dermal pentactines appears to be presently unique in the Palaeozoic fossil record. Mesozoic hexactinellid faunas are dominated by hexactinosan taxa, although unfused skeletons are preserved in the Arnager sponge fauna from Bornholm (Denmark; Brückner 2006); none of her taxa resemble the new species, although several have skeletons combining hexactines, pentactines and diactines.

Hexactinellids represented in the equivalent lithistid-dominated faunas of Arctic Canada are similar in being relatively complex, previously unknown taxa, but none closely resemble *U. euplectelloides* n. sp. Some, such as *Corticulospongia* Rigby & Chatterton, 1989, consist of a similar, thin wall that contains rhabdodiactines, but in that case, these dominate the structure, with abundant interlacing tracts forming a dense layer. *Lumectospongia* Rigby & Chatterton, 1989 is also dominated by monaxons (in this case uncinates) but shares the presence of smooth-rayed hexactines with the new species. Although each represents a significantly more derived structure than normal reticulosans, with features of spiculation resembling *Urphaenomenosoingia*, both these Canadian taxa have a fundamentally different, felted dermal layer. The simpler structure and spiculation of *Malumispongium* (de Freitas 1991) is also fundamentally different.

Recognition of the new species elsewhere may require three-dimensional preservation, as seen here; it is not clear how distinctive the species would be in flattened preservation. However, no previously described species could be reconstructed to produce the architecture described here, and the combination of spicule forms alone is highly unusual in the Palaeozoic; we are therefore satisfied that the species is indeed new and would be recognizable in well-preserved material from elsewhere.

Lyssacinosida? indet.

Plate 21, figures 1–7

Material. – Two specimens on NRM Sp10276, NRM Sp10273 (attached to rip-up clast), NRM Sp10272, NRM Sp10279 and NRM Sp10280.

Description. – Amorphous, bulbous masses of spicules within regions of mostly clear chalcedony, usually attached to spiculitic mud clasts (Pl. 21, figs 3, 6–7) but also associated with a range of skeletal fragments. Each body contains a dense array of spicules. In one specimen (NRM Sp10276; Pl. 21, fig. 1), the sponge appears to have been partly covered by clastic material on at least two occasions, then have regrown at the margins to form a thin sheet over the surface. The upper surface appears at one point to have possessed a sculpture of pronounced, closely spaced depressions, but this may instead represent the surface of a poorly preserved fragment of stromatoporoid with a mounded dermal surface overlying the surface; such sculpture has not been identified in any other specimens. Other specimens (e.g. NRM Sp10280, Pl. 21, fig. 5) are either thin (millimetre-scale) crusts on clast surfaces or were infilling fissures and spaces between larger clasts.

Spicules are diverse, including monaxons, hexactines and apparent anatriaenes (Pl. 21, figs 2, 4, 7), but the spicule complements form a continuum, and spicules are in general substantially largely than those of the other hexactinellids and non-lithistid demosponges described here. In almost all cases, the spicule interior is significantly dissolved, in many cases with only the outer lamina visible; however, a few specimens show irregular axial structures, indicating diagenetically enlarged axial canals. They include large robust hexactines, hexactines with two reduced rays, and probably a substantial component of monaxons. NRM Sp10273 includes anchoring spicules (Pl. 21, fig. 7), with the short, recurved rays of the anchor embedded in a mass of dominantly monaxial rays (perhaps mostly distal rays of hexactines, but some probably also monaxons). These spicules are technically anatriaenes, with three short recurved rays. Spicules in all specimens are arranged irregularly, and in poor specimens (e.g. NRM Sp10272) consist only of densely packed clusters of monaxial or long-rayed spicules up to 0.1 mm in ray diameter.

Largest spicules exceed 2 mm in ray length and have basal ray diameter 0.25 mm; some ray fragments of 2 mm length are only 0.10 mm in diameter, suggesting that the rays of the largest spicules are several millimetres. Rays are smooth (some showing slight siliceous overgrowth or dissolution effects) and straight or very slightly curved. Inter-ray angles appear to be close to orthogonal.

Remarks. – The growth form and spiculation of this sponge are unusual, and probably unique. There are many Palaeozoic hexactinellids with thick walls and irregular internal skeletons (e.g. Brachiospongiidae, Pelicaspongiidae), but these are almost invariably regular in morphology, normally globose, vasiform or digitate. The spicules themselves are not distinctive, except for the unusual anchoring spicule. This type of basalia morphology is known from few groups. Two-spined anchors are present in Palaeozoic dictyosponges (Hall & Clarke 1899), which this sponge clearly is not, and from modern and Mesozoic Pheronematidae (oldest known record is Jurassic; Hinde & Holmes 1892 – although van Kempen 1990b stated that these spicules had four arms). In constrast, four-pronged anchorate hypodermal pentactines appear to be largely or exclusively limited to the Rossellidae (Tabachnik 2002). This group also frequently contains a range of diactines, delicate, long-rayed hexactines, similar to those in these assemblages.

The presence of a three-armed anchoring spicule that is technically an anatriaene shows that these spicules are not limited to demosponges. Anatriaenes have been recorded from Late Ordovician erratics on the island of Sylt, northern Germany, which are probably the oldest known examples (van Kem-

---

Explanation of Plate 20.

Figs 1–4. *Urphaenomenospongia euplectelloides* n. gen. et n. sp.

1–4, NRM Sp10278, holotype; **1**, overall view of well-preserved wall fragment, dermal surface to left; **2**, detail of dermal part of skeleton, showing outer surface with long-shafted orthopentactines (der. pent.), pierced by prostalia hypodermialial rays (hyp. pro.) of large endodermal hexactines; **3**, large-scale view of best-preserved section of wall, showing dermal (d), middle (m) and gastral (g) skeletal layers; the middle layer is composed largely of long-rayed hexactines (l.-r. hex.) combined with probable monaxons; **4**, detail of inner parts of skeletal wall, showing endodermal hexactines (en. hex.) combined with wall-parallel clusters of monaxons on the presumed gastral margin (gast. mon.).

Figs 5–6. Hexactinellid indet. A

5, NRM Sp10286. Dense association of disarticulated spicules, showing robust morphology; **6**, NRM Sp10271. Dense association of hexactines with enlarged axial canals (white); patches of clear chalcedony surrounding spicules possibly indicate articulated preservation with disordered skeletal architecture.

Scale bars represent 1 mm (Figs **2–5**) and 1.5 mm (Figs **1**, **6**).

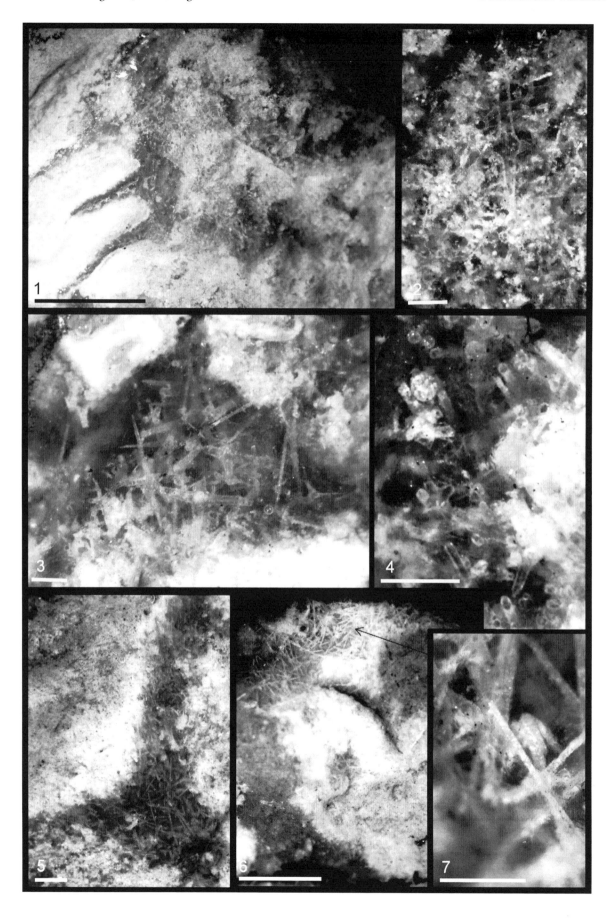

pen 1990b). The examples described here have shorter arms, more closely resembling those of anchorate hexactinellid spicules, but the fundamental morphology is otherwise similar. It is possible that the spicule assemblage in these remains is at least partly agglutinated from detrital spicules, but there is no evidence suggesting this (such as incorporation of numerous broken spicules, or inorganic siliceous particles). It is also possible that this particular specimen does not represent a sponge body, but an inorganic accumulation of spicules, but the overall homogeneity of the spicules, lack of sedimentary grains and apparent position encrusting a large sediment clast all argue against this. The implication is that some anatriaene-like spicules in the Early Palaeozoic may be from hexactinellids, and care must be taken in inferring the occurrence of demosponges from isolated spicules.

Order and family uncertain

Hexactinellid indet. A

Plate 20, figures 5–6

Material. – NRM Sp10286, small pebble with an area containing a dense region of distinctive spicules, distributed in patches within one lamina. NRM Sp10271, small region in another pebble; possibly articulated.

Description. – Small regions of whitish, granular chert containing a large number of large, robust, short-rayed hexactines (Pl. 20, figs 5–6). Spicules are up to 1.2 mm in ray length, with basal ray diameter 0.2 mm, and with largest spicule fragments with basal ray diameter 0.3 mm. Smallest spicules are roughly half the size of the largest, in proportion; smaller spicules may be present, but are not obvious. Rays taper evenly from the centre and appear to be equal and straight, although the semi-opaque silica surrounding them prevents views of all rays in any one spicule. Adjacent areas of matrix in NRM Sp10286 include dense accumulations of fine-rayed spicules.

The second specimen (NRM Sp10271; Pl. 20, fig. 6) may be articulated, with spicules embedded in semi-translucent chalcedony and without admixture of other morphologies. However, there is no clear shape, margin, or structure; spicules are arranged apparently randomly. Spicule proportions as above, with expanded axial canals.

Discussion. – The rarity of similar spicules in the remainder of the samples suggests that this represents, at least in NRM Sp10271, a semi-articulated sponge or a skeleton that has disarticulated within an enclosed region. The spicules are more robust than other hexactines encountered in the fauna and appear to represent a distinct species.

Hexactinellid indet. B

Plate 19, figures 9–10

Material. – NRM Sp10277.

Description. – Single specimen with a semi-articulated fragment that includes semi-aligned orthopentactines or stauractines; no spicules show any evidence for a proximal ray, so stauractines are most likely. Spicules are arranged orthogonally (Pl. 19, fig. 9) and diagonally to each other in the small area that is visible, and are entirely unbundled, but the patterns are obscured by randomly oriented loose spicules. Rays of stauractines are straight, narrow, and gently tapered, with maximum observed ray length 1.0 mm (basal ray diameter approx. 0.03 mm); larger spicules are present, with ray diameter at least 0.05 mm, but are incompletely preserved. Small spicules are not obvious, and where a regular arrangement is visible, it is composed largely of spicules of one size order (ray length approximately 1 mm).

Remarks. – No detailed discussion of this specimen is possible, although the spicules appear to differ from those seen in any other specimen and shows the presence of a different group of hexactinellids in the community. Regular stauractines or dermal

---

Explanation of Plate 21.
Figs 1–7. Lyssacinosida? indet. A
1–3, NRM Sp10276; **1**, overall view of specimen showing multiple regrowth laminae over sedimentary burial layers; **2**, detail of central region showing large hexactines; **3**, second specimen in same pebble, showing encrusting habit onto surface of mud clast. **4**, NRM Sp10279; detail of central region of specimen showing dense monaxon array in clear chalcedony, with small hexactine centres; **5**, NRM Sp10280; dense mass of dominantly monaxons in thin, probably encrusting mat. **6–7**, NRM Sp10273; **6**, overall view of spicular (mostly monaxonid) mass encrusting one surface of a large mud clast, and surrounded by clear chalcedony; **7**, detail (position arrowed in Fig. **6**) showing a single anatriaene-like spicule.
Scale bars represent 1 mm (Figs **2–5, 7**) and 5 mm (Figs **1, 6**).

orthopentactines are present in a variety of early sponge groups, but are dominant in the Reticulosa and some other families such as the Pelicaspongiidae. Most of the very thin-walled taxa such as the Protospongiidae possessed more slender spicules, whereas those of slightly thicker-walled groups tend to be somewhat more robust (Botting 2004). This still includes several plausible superfamilies, and no further discussion would be reasonable. The apparent lack of small spicule size orders is unusual, but occurs in several widely separated genera, implying that the feature is of limited phylogenetic significance.

Spiculites

Plate 22, figures 1–5

Material. – NRM Sp10073, NRM Sp10151, NRM Sp10180, NRM Sp10182 and NRM Sp10284, from Blåhäll/Tofta, comprising dense assemblages of spicules, predominantly monaxons.

Other material. – NRM Sp10247, Sp10248, Sp10249, Sp10252, Sp10253, Sp10254 and Sp10257 are from Blåhäll/Tofta. NRM Sp10255 is from Gnisvärd.

Description. – The pebbles vary in dimensions from $22 \times 19 \times 10$ mm to the largest at $60 \times 37 \times 21$ mm (NRM Sp10151, Pl. 22, fig. 1). The fabric of most is slightly compressed (Pl. 22, fig. 2), confusing the original arrangement of spicules. These monaxons are distinct from much thinner monaxons and hexactinellid spicules that occur commonly as allochthoneous debris in or on the surface of most of the sponge bodies that show adhering matrix. The accumulations of spicules discussed here consist predominantly of oxeas, which are packed so densely as to constitute spiculites. It is notable that such oxeas isolated are very rare or absent in the observed Telychian material. Hexactinellid spicules are distributed randomly throughout some of the spiculite specimens but are not considered further here, although they are useful in confirming the spiculite as part of the Telychian assemblage.

The oxeas are quite uniform in shape and dimensions. They are straight, with the maximum width near the middle. The extremes of the size range are 0.6–2 mm, with width 0.06–0.16 mm, but the dimensions of the overwhelming majority show a narrower range, averaging 1.33 mm long and 0.1 mm wide. In numerous cross-sections, the axial canal is preserved as either powdery matrix or clear chalcedony. Its shape (square, triangular or hexagonal) could not be established. Some specimens, such as NRM Sp10073 and NRM Sp10182, form a coquina with *P. subrectus* (Pl. 22, figs 3, 5). Oxeas in the former specimen are on average 1–2 mm long and 0.12–015 mm thick; they are arranged irregularly, or in small groups of 4–7 diverging oxeas (Pl. 22, fig. 4).

The arrangement of the oxeas is generally sub-parallel, often forming closely packed bundles with little space between spicules. Bundles frequently diverge from an undefined centre into all directions, forming a kind of spherical cluster, about 4–8 mm in diameter. Adjacent clusters sometimes meet distally, and at the junction the arrangement of oxeas is more irregular.

Bundles of sub-parallel, converging monaxons, as occur on the surfaces of several sponge bodies (such as on the stalked part of *Caryoconus gothlandicus*, NRM Sp10065) may be interpreted as root-tufts, preserved more-or-less in original arrangement. Interpretation of such cone-like bundles as monaxonoid sponges seems less likely, as no typically demosponge-like architecture has been observed.

Remarks. – Although the origin of the oxeas in the spiculite is unknown, we regard them as remains of decomposed monaxonid demosponges. In general, the spicules are too small to represent hexactinellid root tuft spicules, which are more typically several millimetres to tens of millimetres in length. We realize that even weak currents during the sedimentation may have caused a selective accumulation of oxeas, but local structures suggest that some original architecture has been preserved. Additional material may reveal the origin of these spiculites.

Atractosella Hinde, 1888 has been reported from Lower Visby Formation (upper Llandovery

---

Explanation of Plate 22.
Figs 1–5. Spiculite and coquina
1, Spiculite; NRM Sp10151. Densely packed monaxons, predominantly oxeas. **2**, Spiculite; NRM Sp10180. The elevated part in the centre is composed of much thinner monaxons with different orientation. **3–4**, Spiculite; NRM Sp10073; **3**, coquina with *Pentameroides* and numerous oxeas; **4**, detail of Figure **3**, showing oxeas that are on average shorter and thicker than usual. **5**, Spiculite; NRM Sp10182. Natural polished surface of a flint-like block. At lower left is a natural cross-section of *Pentameroides*.
Scale bars represent 1 cm (Figs **1–3**, **5**) and 1 mm (Fig. **4**).

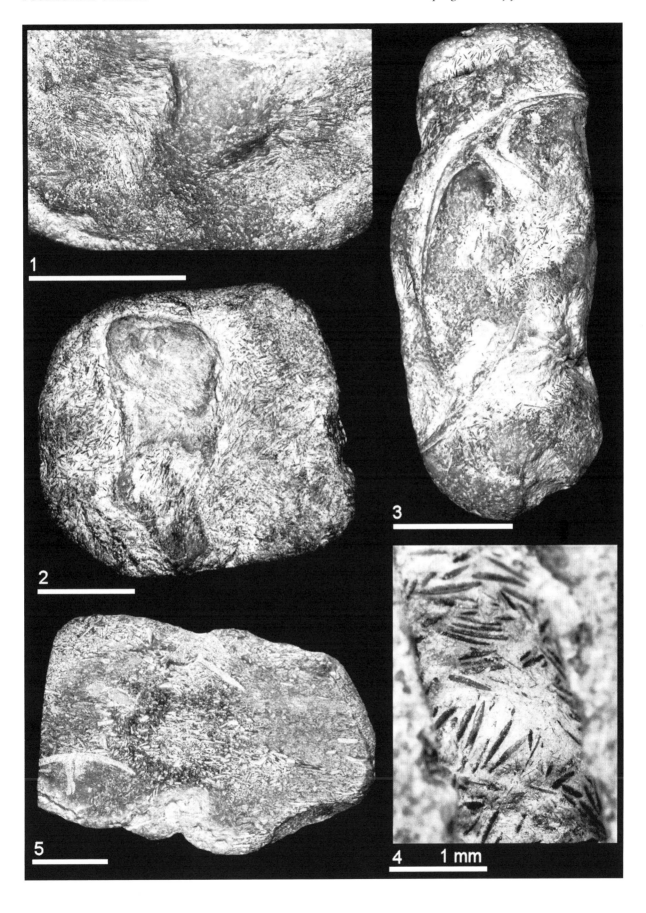

to Wenlock) on Gotland (Bengtson 1979), although Bengtson (1981) has re-identified it as octocoral. Dimensions are considerably longer, and disproportionately thicker than the oxeas described herein; octocoral spicules also lack axial canals.

Class Stromatoporoidea Stearn, Webby, Nestor & Stock, 1999

Remarks. – The systematic position of stromatoporoids has been debated for more than a century. In this study, we follow the view of Stearn *et al.* (1999), who classified Early Palaeozoic stromatoporoids as an extinct group (nominally at class level) of non-spiculate Porifera, represented as fossils by their basal carbonate skeleton. Spicules have been identified in some Middle Palaeozoic and Mesozoic 'stromatoporoids', and those taxa have now been classified within the classes Demospongea and Calcarea. According to Stearn *et al.* (1999), the class Stromatoporoide and the term 'stromatoporoid' should refer only to corraline sponges of Ordovician to Devonian age.

Stromatoporoids probably occur more frequently in this sponge assemblage than is currently recognized. Many specimens are difficult or impossible to distinguish from specimens from younger strata, such as the Högklint Formation, or from erratics with unknown provenance of Late Ordovician age. The only stromatoporoid specimens that have been selected for description are those associated with one of the sponges of demonstrable Telychian age in the same block. This restrictive criterion resulted in only five specimens. Three associations of a stromatoporoid with a demosponge occur: with *Caryoconus* Rhebergen & van Kempen, 2002, *Somersetella* Rigby & Dixon, 1979 and *Haplistion* Young & Young 1877, respectively.

Order Labechiida Kühn, 1927
Family Stylostromatidae Webby, 1993

Genus *Pachystylostroma* Nestor, 1964

Pachystylostroma sp.

Plate 23, figures 1–2

Material. – NRM Sp10175; collected from Blåhäll/ Tofta.

Description. – The skeleton of *Pachystylostroma* sp. is mammilate and contains cyst plates of variable size alternating with lamina-like layers. Mamelon columns may have simple upward-and-outward branching pillars (Nestor 1964). The sub-hemispherical, greyish coenosteum of the only specimen yet recognized is part of a smoothly worn block, NRM Sp10175, from Blåhäll/Tofta (Pl. 23, fig. 1). The base is incomplete, but the upper surface is well preserved. It is covered by reddish-brown matrix, parts of which consist of translucent chalcedony encompassing fossil fragments, among them locally abundant hexactinellid spicules.

The coenosteum is irregularly laminar and undulating, but is often interrupted by mamelon columns, 4–6 mm apart. Some of these columns bifurcate upwards at their base. There are 4–7 equal-sized pillars, about 2.0 mm long and 0.10 mm wide, in the basal and middle parts of most of the columns. Pillars in the basal part tend to converge, whereas those in the middle part are parallel, or diverge from the mamelons. Laminae are about 0.06–0.12 mm thick and separate cyst plates, which are on average 0.10 mm high. Cysts generally have a convex upper surface, but some are concave. Numerous well-developed denticles occur on both cysts and plates. To accentuate the undulating coenosteum, the counterpart of NRM Sp10175 has been figured from the top (Pl. 23, fig. 2).

Explanation of Plate 23.
Figs 1–6. Stromatoporoids
Stromaporoids are attached to a demosponge or with hexactinellid spicules.
Figs 1–2. Pachystylostroma sp. Nestor, 1964
1–2, NRM Sp10175; **1**, vertical section. Numerous hexactinellid spicules occur in the dark-coloured chalcedony at left; **2**, counterpart of Figure 1, in top view (see text).
Fig. 3. Clathrodictyon sp. Nicholson & Murie, 1878
3, NRM Sp10141. Arrow indicates a small part of Haplistion sp.
Fig. 4. Syringostromella Nestor, 1966
4, NRM Sp10150. The stromatoporoid is attached to a fragment of Caryoconus gothlandicus and surface matrix contains numerous isolated spicules.
Figs 5–6. Amphipora sp. Schulz, 1883
5–6, NRM Sp10085; **5**, weathered lateral surface; **6**, polished surface, showing the digitate stromatoporoid, and the attached Somersetella sp. Scale bars represent 1 cm.

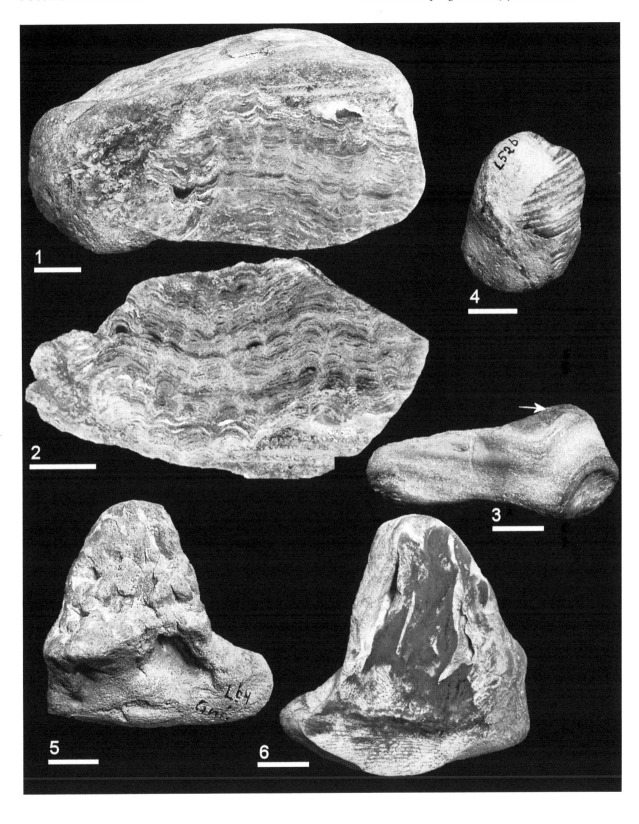

Remarks. – A small tabulate, probably *Favosites* sp., settled on the upper surface of the coenosteum and was overgrown by it (it is white colouring contrasts with the stromatoporoid in the upper-right in (Pl. 23, fig. 1). Isolated hexactinellid spicules occur throughout the matrix, but are concentrated in areas of translucent chalcedony (Pl. 23, fig. 1, at left).

Mori (1969) erected the species *P. visbyensis*, based on a single specimen from the Visby Beds near Nyhamn, Gotland. A direct comparison with that species was not possible, as micro-structures could not be studied in the present specimen, due to intensive replacement by chalcedony. Therefore, we leave the assignment to *Pachystylostroma* on generic level.

Order Clathrodictyida Bogoyavlenskaya, 1969
Family Clathrodictyidae Kühn, 1939

Genus *Clathrodictyon* Nicholson & Murie, 1878

Clathrodictyon sp.

Plate 23, figure 3

Material. – NRM Sp10105 and NRM Sp10141; from Blåhäll/Tofta. NRM Sp10129; from Gnisvärd.

Description. – NRM Sp10141 is a small, abraded specimen, coloured a combination of reddish and grey/white. Tangentially, it preserves regularly arranged parallel laminae (Pl. 23, fig. 3). It has a concave base and a convex upper surface, on which the rhizoclonid *Haplistion* sp. has nucleated (see arrow in Pl. 23, fig. 3). The sponge fragment is recognizable by tracts that enclose canals, but the sponge is too poorly preserved to recognize individual rhizoclones.

Remarks. – NRM Sp10105 is part of a coquina, comprising *Subalveolitella* sp., *Clathrodictyon* sp., as well as numerous hexactinellid spicules.

Order Stromatoporida Stearn, 1980
Family Syringostromellidae Stearn, 1980

Genus *Syringostromella* Nestor, 1966

Syringostromella sp.

Plate 23, figure 4

Material. – NRM Sp10150; collected from Blåhäll/Tofta.

Description. – A single fragment of *Syringostromella* sp. has been preserved in NRM Sp10150. The small pebble, collected from Blåhäll/Tofta, is associated with *Caryoconus gothlandicus*. They are separated from each other by a thin zone of silicified limestone with fragments of crinoid columnals and numerous hexactinellid spicules (Pl. 23, fig. 4).

The coenosteum is laminar. Laminae were probably denser than intervening areas and were more heavily silicified, resulting in them being more resistant to erosion. In lateral view, numerous small, vertical pillars form an elaborate reticulate mesh with horizontal laminae and interlaminae plates. Pillars in tangential section form an irregular, vermiculate and anastomosing network.

Remarks. – Mori (1969) described some species of *Syringostromella* from Wenlockian strata of Gotland. According to Stearn *et al.* (1999), the first appearance of *Syringostromella* was in the Lower Silurian (Telychian), from the Hudson Bay lowlands (Canada) and the Oslo region (Norway). The present material is too limited, and too poorly preserved, to identify at species level.

Order Amphiporida Rukhin, 1938
Family Amphiporidae Rukhin, 1938

Genus *Amphipora* Schulz, 1883

Amphipora sp.

Plate 23, figures 5–6

Material. – NRM Sp10085; collected from Gnisvärd.

Description. – *Amphipora* Schulz, 1883 is a dendritic, dichotomously branching stromatoporoid with pillars that radiate upward and outward from the axis. NRM Sp10085 from Gnisvärd beach is the only digitate stromatoporoid in the assemblage. The specimen is coloured deep red through oxidation; the surface is intensively weathered, as shown in Plate 23, figure 5. It is attached perpendicularly to an anthaspidellid sponge, most likely *Somersetella* sp. Rigby & Dixon, 1979 (Pl. 23, figs 5–6). The specimen has been cut laterally.

Remarks. – According to Stearn *et al.* (1999), *Amphipora* is a common stromatoporoid in the Devonian, but has also been recorded from Ludlovian, and possibly fromWenlockian strata; this occurrence, if confirmed, would therefore represent the oldest record. Its identification, based on the limited material, should be considered tentative, but the unusual combination of a digitate stromatoporoid attached to a rhizomorine sponge justifies a brief description.

Palaeobiogeography and community evolution

Very few sponge assemblages of Telychian age have been recorded worldwide (Muir *et al.* in press). The only Llandovery-age lithistids described to date are *Calycocoelia* (Zhan-Qiu 1990) from the Yangtze Platform, and *H. sphaeroidalis* (Rhebergen 2005). Some comparable associations are considerably older, of Late Ordovician age, such as those from northwestern Europe, North America and New South Wales (Australia). Other similar faunas are significantly younger, such as the assemblages described from Arctic and northwest Canada, which are of Wenlock to Ludlow age (see below). To our knowledge, there are no published reports of articulated sponges of Rhuddanian or Aeronian age, from Baltica or elsewhere.

To place the new fauna in context with other sponge associations, we have compared the present one with preceding (Late Ordovician) and succeeding (Wenlock to Ludlow) faunas:

1 Upper Ordovician assemblage from the Malongulli Formation, New South Wales, Australia (Rigby & Webby 1988);
2 Upper Ordovician assemblage of erratic sponges on Gotland (Rhebergen & von Hacht 2000) from as yet unknown provenance on Baltica;
3 Silurian (Wenlock to Ludlow) assemblages from northern Canada (Rigby & Dixon 1979; Rigby & Chatterton 1989, 1999; de Freitas 1989, 1991).

Comparison with Upper Ordovician sponges from the Malongulli Formation

The assemblage from the Upper Ordovician Malongulli Formation from New South Wales, Australia, has a strongly endemic character. Of 34 genera listed, only four taxa are shared with the Gotand assemblage: the rhizomorine genera *Haplistion* and *Warrigalia*, and the cosmopolitan species *Archaeoscyphia minganensis* and *H. sphaeroidalis*. Remarkably, more taxa are shared between the Malongulli association and the Arctic Canada associations (see below), perhaps suggesting that many rare, shared

taxa are yet to be recovered from the Gotland assemblage.

Comparison with Late Ordovician erratic sponges on Gotland

Although the Ordovician and Silurian sponge assemblages on Gotland are intermixed, their taxonomic compositions show conspicuous differences. The Ordovician assemblage from Baltica consists almost exclusively of lithistid demosponges of the suborder Orchocladina, with five families represented: Anthaspidellidae, Streptosolenidae, Astylospongiidae, Chiastoclonellidae and Hindiidae. The only representative of the Hexactinellida recorded is 'Pyritonema' sp. (M'Coy, 1850; Rhebergen *et al.* 2001), which most probably should be re-interpreted as an indeterminate root-tuft. Recently, Botting & Rhebergen (2011) described *Haljalaspongia inaudita*, collected in the westernmost part of Germany; this is the first hexactinellid sponge known from the Ordovician (Sandbian; Upper Ordovician) erratic blocks of silicified limestone originating in Baltic region.

The large majority of taxa in the Telychian assemblage are absent from the Ordovician assemblage. The assemblages have only six species in common, two of which are cosmopolitan (*H. sphaeroidalis* and *Archaeoscyphia minganensis*). The differences become even more apparent on generic level; the most frequent Ordovician genera, *Aulocopium*, *Astylospongia*, *Carpospongia* and *Caryospongia*, comprising more than 86 per cent of the specimens, are absent from the Silurian assemblage.

In contrast, the Silurian assemblage includes, in addition to the families in the Orchocladina, several genera of Rhizomorina, numerous spicules of hexactinellid sponges (of at least four species, and probably many more), as well as non-lithistid demosponges. The most frequent Silurian genus, *Caryoconus* (Rhebergen & van Kempen, 2002), comprising about 50 per cent of the assemblage, is apparently endemic (although it possibly occurs in Arctic Canada, as discussed in Systematic Palaeontology). The genus *Archaeoscyphia* is the second most frequent genus in the Telychian assemblage and is represented by seven species, but it is rare in the Ordovician Baltic assemblages and represented by only one or possibly two species.

These differences lead to the conclusion that the present assemblage does not represent a direct succession from the Ordovician fauna of the same region, but results from repopulation of the environment following a major extinction of sponges during the Late Ordovician biotic crises.

Comparison with Silurian assemblages in northern Canada

Diverse Silurian sponge assemblages have been described from Arctic and northwest Canada. Rigby & Dixon (1979) described a sponge fauna from the Upper Silurian (probably Ludlovian) Read Bay Formation, Somerset Island (Arctic Canada). Rigby & Chatterton (1989) described somewhat older (Wenlock and Ludlow) sponge associations from Baillie–Hamilton and Cornwallis Islands (Arctic Canada). De Freitas (1989) described orchocladinid sponges of Wenlockian age, as well as a more varied sponge fauna consisting of lithistids and hexactinellids of Ludlovian age (de Freitas 1991), both from Cornwallis Island. Rigby & Chatterton (1994, 1999) described a hexactinellid sponge and an assemblage of Wenlockian demosponges from the Mackenzie Mountains (Northwest Territories, Canada).

The striking similarities between the present assemblage and those from northern Canada encourage a more detailed comparison. Note that the hexactinellids in the Gotland assemblage are with two exceptions (unique to the assemblage) poorly known and are therefore left out of the discussion; hexactinellids are much more easily destroyed biostratinomically than lithistids. It is likely that the non-rigid skeletons of these Silurian hexactinellids were in most cases lost through disarticulation, and they may originally have constituted a much greater proportion of the community; the preserved assemblages may therefore be random samples of the same assemblage.

The lithistids mentioned in the assemblages discussed here are listed in Table 3. There is a striking similarity in the composition of occurring taxa: fourteen or fifteen of 22 species in the Gotland assemblage also occur in Canadian assemblages, which total 48 or 49; only seven or eight species are not known from Canada. The uncertainty relates to two specimens in the sponge assemblage from the Mackenzie Mountains (northwest Canada), which Rigby & Chatterton (1999) assigned tentatively to *Astylospongia praemorsa*. These specimens may need to be reassigned to *Caryoconus* Rhebergen & van Kempen, 2002), but examination of the internal canal system is needed.

All assemblages known from Northern Canada are considered to be significantly younger than the Telychian sponge association from Baltica. At first sight, these different ages may lead to the assumption of a migration from Baltica to Laurentia, but that is as yet premature. Further investigation and comparison of the associations is needed and may yield new information on palaeobiogeographical and palaeo-

stratigraphical aspects of the distribution of Early Silurian sponges. More critically, additional faunas are needed to establish palaeobiogeographical patterns in Silurian sponges (Muir *et al.* in press), particularly from the Llandovery. Strong faunal overturn during the end-Ordovician interval, followed by the apparent homogeneity of faunas through the Silurian, across Baltica and Laurentia, suggests that a small number of taxa survived the extinction event, diversified and subsequently dispersed across at least some low latitude regions.

The Ordovician sponge association from Baltica probably disappeared during the Hirnantian (end-Ordovician) extinction events. Equatorial taxa may have been particularly sensitive to a cooling event, as the temperature regime to which they were adapted no longer existed, whereas taxa from higher latitudes may have survived by shifting to lower latitudes. It would be interesting to see whether the Silurian assemblage from Gotland is more similar to higher-latitude Late Ordovician assemblages, but the data necessary for comparison are not yet known. Both *Aulocopium* and *Archaeoscyphia* are present in the Middle Ordovician San Juan Formation of the Argentine Precordillera (Beresi & Rigby 1993), but this fauna is too old for useful comparison with the Telychian assemblage.

Aldridge *et al.* (1993) and Jeppsson (1998) summarized general characteristics of several events related to sea-level changes during the Llandoverian. In particular, their correlations of the chronostratigraphy, oceanic regime and sea-level curves with the standard conodont and graptolite biostratigraphy are important for this study, and their scheme has been reproduced in adapted form in Figure 11 (stratigraphy after Jeppsson 1998). The sponge assemblage lived during the Snipklint Primo Episode, the relevant period in this study. The *M. spiralis* Graptolite Biozone, with which the sponge fauna can be correlated, appears to be coeval with the *P. amorphognathoides* Conodont Biozone. Although graptolites and conodonts have not yet been recorded, the age has been established on account of associated acritarchs (see geological setting and materials and methods sections, above).

Aldridge *et al.* (1993) characterized the Snipklint Primo Episode as an interval with increased humidity, which led to increased weathering and correspondingly increased deposition of argillaceous sediment; the interval also yields abundant and diverse communities, which lived in well-oxygenated ocean waters. This allowed 'infaunal and epifaunal benthic organisms to survive in places where they were formerly rare, due to low levels of oxygen and/or food' (Aldridge *et al.* 1993, p. 508). They

Table 3. Comparison of Silurian demosponge assemblages from Gotland and northern Canada.

Locality Age Taxa	Gotland Llandovery	NW Canada Wenlock	Cornwallis Island Wenlock–Ludlow	Baillie-Hamilton Island Wenlock–Ludlow	Somerset Island Ludlow
Rhizomorina					
Haplistion sp. Young & Young, 1877	x			x	
Haplistion minutum Rigby & Dixon, 1979	xx				x
Haplistion cylindricum Rigby & Dixon, 1979	x				x
Haplistion creswelli Rigby & Dixon, 1979					x
Haplistion frustrum Rigby & Chatterton, 1989		x		x	
Haplistion toftanum n. sp.	x				
Warrigalia robusta Rigby & Webby, 1988	xx				
Parodospongia euhydra Rigby & Chatterton, 1989				x	
Megamorina					
Eochaunactis radiata Rigby & Dixon, 1979					x
Haplistionella garnieri Rigby & Dixon, 1979					x
Haplistionella minitraba Rigby & Dixon, 1979					x
Orchocladina					
Archaeoscyphia annulata (Rigby, 1973)	xx		xx		
Archaeoscyphia aulocopiformis de Freitas, 1989	xx		x		
Archaeoscyphia attenuata de Freitas, 1989	x	xx	xxx		
Archaeoscyphia alternata de Freitas, 1989		x	xx		
Archaeoscyphia gislei de Freitas, 1989	xxx	xxx	x		
Archaeoscyphia minganensis (Billings, 1859)	xxx		xx	x	
Archaeoscyphia rectilinearis de Freitas, 1989	xx	xxx	x		
Archaeoscyphia scalaria de Freitas, 1989	xx	x	xxx		
Antrospongia aberrans Rigby & Chatterton, 1989				x	
Aulocopium nana Rigby & Chatterton, 1989				x	
Calycocoelia typicalis Bassler, 1927	x				
Calycocoelia micropora Rigby & Chatterton, 1989			x		
Dunhillia spp. Rigby & Webby, 1988			x	x	
Patellispongia spp. Bassler, 1941		xxx			
Finksella turbinata Rigby & Dixon, 1979	xx				x
Somersetella conicula Rigby & Dixon, 1979			xxxx		xx
Somersetella amplia Rigby & Dixon, 1979	x		x		
Somersetella digitata Rigby & Dixon, 1979	x	x	x		
Climacospongia undulata de Freitas, 1991	xx		x		
Climacospongia snowblindella de Freitas, 1991			x		
Rhodesispongia simplex de Freitas, 1991			x		
Multistella leipnitzae n. sp.	xxx				
Postperissocoelia gnisvardensis n. gen., n. sp.	x				
Postperissocoelia spinosa (Rigby & Chatterton, 1989)			x		
Perissocoelia gelasinina Rigby & Chatterton, 1989				x	
Chiastoclonella sp. Rauff, 1895	x				
Astylospongia praemorsa? (Goldfuss, 1826)		x ?			
Astylospongiella (?) *lutera* Rigby & Chatterton, 1989			x		
Astylospongiella megale Rigby & Lenz, 1978				x	
Astylospongiella striola Rigby & Chatterton, 1989			x	x	
Caryospongia juglans Quenstedt, 1878			x		
C. tuberosa de Freitas, 1991			xx		
Caryoconus gothlandicus (Schlüter, 1884)	xxxx				
Lindstroemispongia cylindrata n. gen., n. sp.	x				
Palaeomanon cratera (Roemer, 1848)		x			
Carpospongia castanea (Roemer, 1861)		x			
C. globosa (Eichwald, 1830)				x	
Hindia sphaeroidalis Duncan, 1879	xxx	xxx	xxxx	xxx	x

x = 1–5; xx = 5–10; xxx ≥ 10; xxxx = numerous.

considered this episode to indicate 'the first return of full cycling of nutrients after the Ordovician glaciation, producing bonanza conditions, especially for plankton and plankton-dependant communities' (Aldridge *et al.* 1993, p. 509). Sponges would have been directly encouraged by increases in plankton abundance due to their filter-feeding habit and probably underwent rapid expansion of faunal ranges and abundance, and perhaps also diversification.

Some cosmopolitan species colonized the empty ecospace, in combination with new forms that radi-

Chrono-stratigraphy		Oceanic Regime	Standard Conodont Zonation	Graptolite Zonation	Sea-level Curve
					Low High
W E N L O C K	GLEE-DON	Klinte Secundo Episode	O. bohemica Zone	C.? gerhardi / C.? ludensis	
				C.? deubeli C.? praedeubeli G. nassa P. d. parvus	
		Mulde Secundo Secundo Event			
	WHIT-WELL	Hellvi Secundo Episode	O. s. sagitta Zone	C. lundgreni	
		Valleviken Ev.			
	S H E I N W O O D I A N	Allekvia Primo Episode	K. o. ortus Zone	C. perneri	
		Lansa Secundo Episode	post K. walliseri interregnum		
		Boge Event	uppermost K. w. range		
		Sanda Primo Episode	K. patula Zone	C. rigidus	
		Vattenfallet Secundo Episode	Middle K. walliseri Zone	M. belophorus	
			Lower K. walliseri Zone		
			O. s. rhenana Z.	M. antennularius	
			Upper K. ranuliformis Z.	M. riccartonensis	
		Ireviken Event	5 Zones	3 Zones	
L L A N D O V E R Y	TELY-CHIAN	Snipklint Primo Episode	P. amorphognathoides Z.	C. lapworthi M. spiralis M. crenulata Mcl. griestonien. M. crispus	
			P. celloni Zone	St. turriculatus St. guerichi	
		Sec.-Pri. Event			
	AERO-NIAN	Malmøykalven Secundo Episode	Distomodus staurognathoides Zone	St. sedgwickii	
		Sandvika Event			
		Jong Primo Ep.		Dem. convolutus 3 Zones	
	RUDDA-NIAN	Spirodden Secundo Episode	Distomodus kentucky-ensis Zone	C. cyphus Cyst. vesiculosus Par. acuminatus	
		?-Sec. Event		G. persculptus	

Fig. 11. Early Silurian oceanic changes (after Jeppsson 1998). The arrow indicates the Snipklint Primo Episode, coeval with the *Monograptus spiralis* Graptolite Biozone and the sponge fauna.

ated within a geographically restricted area. This may have occurred during a short period in the Late Telychian, when conditions ameliorated, but given the poor fossil record it is possible that the fauna had been evolving steadily since the end of the Ordovician. Rapid *in situ* diversification would explain the relatively large number of new taxa, and the endemic characteristics that are quite distinct from the local Ordovician community. It is possible, however, that the fauna diversified elsewhere and spread into Baltica as a coherent sponge community, as appears to have happened with the spread of the Baltic assemblage into Arctic Canada. Modern planktonic sponge larvae usually settle within a few days, a trait that hampers rapid dispersal. If the Got-land fauna migrated in from elsewhere, it is likely that this was not a rapid process, and therefore if the Late Telychian climatic conditions were important, the fauna probably evolved within the local area.

Taphonomy

Three aspects of the diagenesis of the sponges deserve brief discussion. Firstly, most of the cylindrical sponge bodies have been compressed laterally, sometimes to such a degree that spongocoels with a diameter up to 20–40 mm have been compressed to a slit. However, among the roughly 1000 specimens of the astylospongiid *Caryoconus*, fewer than five specimens show any compression, and where present it is slight. This difference is caused by the framework of the skeleton: spheroclones in astylospongiids were firmly fused, making the sponge body stiff and resistant to compression. In contrast, dendroclones in the ladderlike anthaspidellid structure were connected, but not fused, allowing the sponge body a certain degree of flexibility, and as a result they were easily compressed during compaction of the sediment.

Secondly, some of the tubular spongocoels of the anthaspidellids were filled with matrix. However, the spongocoels of many sponges embedded horizontally were filled with matrix in the lower part only, whereas the upper part remained as a cavity. During diagenesis, these cavities were filled with chalcedony, usually concentrically layered as in agate (Pl. 3, fig. 5). This implies that the chalcedony formed diagenetically early, as otherwise the void areas should have been compressed in most taxa.

Finally, this early silicification appears in some cases to have been extremely rapid. In the hexactinellid *U. euplectelloides* n. gen., n. sp., this transparent chalcedony surrounds a three-dimensional, articulated array of unfused spicules, which could only have been preserved prior to soft tissue decay. Translucent, slightly milky chalcedony forms discrete patches between the spicule layers, suggesting a distinction between soft tissue replacement and infilling of pore or canal space; similar material also infills canals in the best-preserved lithistid specimens, with a distinct boundary against the clear chalcedony that immediately surrounds the skeletal architecture. This appears to represent a previously unrecognized form of soft-tissue preservation. However, it may be restricted to sponges and be dependent on a high proportion of silica in the form of micro-scleres within the sponge tissue.

Palaeoecology and palaeoenvironmental interpretation

Specific fossil associations in the community are listed in Table 2. The combination of fine sediment substrate with a faunal assemblage including abundant sponges, relatively few stromatoporoids and tabulates, and only rare rugose corals, suggests a community in an outer platform area, with a relatively high sedimentation rate, but a rather firm, muddy seafloor. Disarticulation of hexactinellid and monaxonid sponges, together with shell coquina development, indicates at least occasional high-energy conditions, probably with near-constant agi-

tation. The frequent monospecific occurrence of the brachiopod *P. subrectus* supports this conclusion, based on the environmental preferences described by Jin & Copper (2000) in their study of Late Ordovician and Silurian pentamerid brachiopods from Anticosti Island. *P. subrectus* is listed as one of the latest species recorded, confined to the Middle and Late Telychian, that is, the Pavillon Member of the Upper Jupiter Formation. The fauna of this mudstone unit is assigned to the Benthic Assemblage scale 4–5 (Boucot 1975). According to Johnson *et al.* (1991), coeval mudstones in Estonia correspond to a depth of BA 5, suggesting an absolute depth of about 50 m (Brett *et al.* 1993); this is in accordance with our interpretation of the depth of the Gotland assemblage.

Narbonne & Dixon (1984) described probably Ludlovian reefs on Somerset Island (Arctic Canada), including a sponge association (Rigby & Dixon 1979) that is comparable with the Telychian assemblage from Gotland. According to Narbonne & Dixon (1984), the reefs illustrate a consistent vertical zonation of lithology and fossil content. The basal 'Crinoid Stage' consisted mainly of crinoid debris, with an abrupt transition to the 'Sponge Stage', composed of mudstone and approximately 80 per cent by volume of lithistid sponges. In the upper part, the number of stromatoporoids, tabulates and corals increased, gradually developing into the third 'Coral Stage', in which these groups became predominant and the volume of sponges decreased considerably and eventually disappeared. These reefal mudmounds, some metres in height and 5–35 m in diameter, were surrounded by a halo of debris, in which the zonal structure of the reef was mirrored. According to Narbonne & Dixon (1984), the combination of very fine-grained matrix and fragile reef-builders indicated an environment of quiet water conditions, well below storm wave-base and with high turbidity. However, the reef-building organisms listed are capable of withstanding highly turbulent conditions, and a shallower water depth seems more likely.

The palaeoecology of the Telychian assemblage of Gotland may have followed a similar pattern, although the associated faunas do not support a directly analogous palaeoecology. Among several hundred fossils with adhering sediment, crinoids are limited to a single holdfast and fewer than ten columnals. Bryozoans are represented only by five specimens of a cryptostome species. Trepostome bryozoans, trilobites, molluscs and algae have not yet been recorded. This composition is in accordance with the 'Sponge stage' as described above, but the absence of crinoid debris suggests that if the community was equivalent, it was not part of a progressive sequence in the development of mudmound architecture.

Although the bedrock strata are not exposed on Gotland, there are similarities between the available loose blocks and the reefs described above. Both associations share the combination of very fine-grained matrix and an abundance of fragile hexactinellid sponges. Assuming the sponges to be derived from the 'Sponge Stage', it may explain the paucity of crinoids and bryozoans from the basal 'Crinoid Stage', as well as the rare association of sponges with tabulates and stromatoporoids, as representatives of the 'Coral Stage'. However, this presupposes that the source area for these blocks was restricted only to one stage of the mound development or that the rest of the deposit was not preserved in the pebble deposits. The latter scenario is possible in the light of the rapid silicification necessary to explain the taphonomy, discussed above. It is possible that the silicification affected only the part of the mound with abundant sponge bodies and spicules, which acted as both a silica source and nucleation. If that is the case, then the carbonate and mud-dominated parts of the mound sequence could have been destroyed during transport, or during one of the weathering stages.

Although no direct evidence of storm-generated structures could be observed, there are three lines of evidence, indicating substantial water movement. First, the numerous hexactinellid sponges were fully disarticulated except for rare articulated fragments, and isolated spicules have been transported and deposited in sheltered areas adjacent to sponge bodies, particularly in spongocoels, superficial grooves and constrictions. Secondly, almost all of the (sub-) cylindrical anthaspidellids, such as *Archaeoscyphia* spp., and the cylindrical rhizomorines, such as *H. cylindricum* and *H. toftanum* n. sp., have been preserved horizontally, whereas sponges embedded in quiet-water mud mounds should have been preserved vertically in many cases. Third, rare coquinas composed of fragments of brachiopods and sponges indicate winnowing and transport during episodes of sustained high energy. The lack of complete hexactinellid sponges also suggests that conditions were not quiet with occasional rapid burial, but rather were consistently agitated. This is indicative of conditions at or above normal wave base, perhaps with storm-influenced rapid sedimentation episodes, but with generally turbulent conditions.

Several aspects of autecology and synecology can also be investigated in the fauna, despite the lack of

bedrock exposure. The frequent occurrence of *P. subrectus* Schuchert & Cooper, 1931 allows recognition of host preference. Association of *Pentameroides* with *Caryoconus* (Pl. 12, fig. 2) is about ten times more frequent than it is on anthaspidellids, and as yet has not been seen with *Hindia* or rhizomorine sponges such as *Haplistion*. The brachiopod in most cases presumably settled on a dead sponge body, using it as substrate, but some brachiopods seem to have settled on living sponge bodies, as the margins of the shell appear to have been overgrown by the sponge skeleton (Pl. 5, fig. 9; Pl. 8, fig. 3). There are even many examples of overgrowth to such a degree that the brachiopod is recognizable only in cross-sections of valves or septa (Fig. 12).

In contrast to the successful Early Ordovician radiation of the anthaspidellids, which apparently adapted rapidly to habitats in a range of depths from deep slope conditions to shallow water, the short duration of the Telychian sponge assemblage implies they had only limited potential for widespread colonization, or very specific environmental requirements. This limitation, whatever its cause, prevented them from becoming common over wide areas, although many genera survived through the Baltica–Laurentia region as a whole for most of the Silurian. It is possible that they could not compete effectively

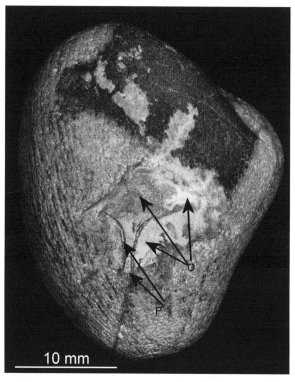

Fig. 12. Example of overgrowth of the brachiopod *Pentameroides subrectus* by an orchocladinid, that is, *Archaeoscyphia gislei* (NRM Sp10204).

with more rapidly growing or physically more resistant organisms such as tabulates and stromatoporoids (Narbonne & Dixon 1984; Johns 1994).

Conclusions

The sponge fauna described here is by far the most diverse fauna yet recorded from the Llandoverian worldwide and provides our best evidence yet for the evolutionary history of sponges following the end-Ordovician extinction episodes. The lithistid-dominated fauna is very different to the pre-existing Late Ordovician assemblage in the same area, implying a dramatic faunal overturn. In contrast, there is a striking similarity of the fauna to that seen in later Silurian deposits in Laurentia, indicating that the re-established sponge faunas rapidly became widespread and successful.

Despite its diversity, the sponge assemblage described here disappeared or became extinct during the Late Telychian in this part of Baltica. Sponges are absent or have not been preserved in the overlying Visby Formation, apart from some rare specimens of *H. sphaeroidalis* in the Lower and/or Upper Visby Formation (Rhebergen 2005) that indicate that universal dissolution of skeletons was not responsible for their absence. It is possible that future examination of the 'Red Layer' may answer some questions, such as the relationship between the local extinction of the sponge assemblage and severe environmental perturbations, possibly including chemical changes in the water caused by volcanic ash input (Laufeld & Jeppsson 1976). In contrast to the fate of the siliceous sponges, coeval assemblages of tabulates, rugose corals and stromatoporoids continued through the Visby formations. This suggests that the cause of their decline may have been related to chemical factors (e.g. levels of dissolved silica), or physical factors (e.g. a switch to higher-energy conditions) that affected spiculate sponges more than other groups. The apparent reliance of the sponges on particular stages of reef-mound development may also be significant.

The fauna of non-lithistid demosponges and hexactinellids is also unusual. These taxa are rarely preserved in shallow-water, high-energy situations due to fragmentation of their skeletons, and so we have little understanding of their evolutionary history in this type of environment. The fauna described here includes rare preservation of delicate skeletal architecture, and even soft-tissue replacements by silica, and reveal a fauna that more closely resemble described Mesozoic assemblages than Palaeozoic ones. As discussed by Botting (2005), this suggests that shallow-water environments were critical to the evolution of derived morphologies and architectures in hexactinellids and non-lithistid demosponges, but their record is largely hidden by taphonomic biases. This fauna shows that even in turbulent environments such taxa can be preserved and further encourages the search for similar faunas in the Late Ordovician – Llandoverian interval, which appears to have been critical to the development of modern lineages.

Acknowledgements. – We are grateful to Mrs. Heilwig Leipnitz, Uelzen (Germany), for much discussion and support, and the loan of hundreds of sponges, as well as for the initial information about the sponge collection in Visby. This study would have been impossible without her indefatigable assiduity in collecting on Gotland. We are indebted to Christina Franzén and her staff and to the Late Valdar Jaanusson in the Department of Palaeozoology of the National Museum of Natural History (Stockholm), as well as to Sara Eliason in the County Museum of Gotland (Visby, Sweden), both for loan of material, valuable discussions and relevant information. Lennart Jeppsson contributed importantly by extended discussions and commenting on earlier drafts. Lars Karis and Linda Wickström (SGU, Uppsala) provided valuable data on the occurrence of the 'Red Layer'. FR is also grateful to his late friend, Ulrich von Hacht, Hamburg, for cooperation, advises and inspiring discussions during many years. Zwier Smeenk (former Laboratory of Palaeobotany and Palynology, University of Utrecht, NL) did the indispensable acritarch research and examined residues of tens of dissolved samples, thus documenting the Silurian age of the sponges. FR thanks Harry Huisman (curator of the former Natuurmuseum Groningen NL), for identifying tabulates and stromatoporoids, as well as Jisuo Jin (Department of Earth Sciences, University of Western Ontario, London, Canada), Madis Rubel (Geological Institute of the Technical University Tallinn, Estonia) for help in identifying brachiopod taxa. FR is also indebted to Andrzej Pisera (Institute of Paleobiology of the Polish Academy of Sciences, Warszawa, Poland) for loan of some specimens from Arctic Canada, advise and helpful discussions. We also thank Peter and Karin de Vries (Sappemeer, NL) for unlimited access to their collections and for loan of some important specimens, Saskia Kars (Institute of Earth Sciences, Free University, Amsterdam) for scanning and photo editing, as well as her colleague, Wynanda Koot, for cutting and preparing a number of slabs, Gerard Beersma (curator former Ecodrome, Zwolle NL) for help with photography, Gerrit Anninga (†) for cutting tens of sponges and Carien Hut (Emmen, NL) for preparation of one of the specimens. We appreciate the helpful comments of reviewer Ronald Johns, and the editor Svend Stouge, both of whom have helped to improve the manuscript. JPB is funded by the Chinese Academy of Sciences Fellowships for Young International Scientists Grant No. 2010Y2ZA03 and National Science Foundation of China, The Research Fellowship for International Young Scientists (Grant No. 41150110152).

References

Aldridge, R.J., Jeppsson, L. & Dorning, K.J. 1993: Early Silurian oceanic episodes and events. *Journal of the Geological Society, London* 150, 501–513.

Bassler, R.S. 1927: A new early Ordovician sponge fauna. *Journal of the Washington Academy of Science* 17, 390–394.

Bassler, R.S. 1941: The Nevada Early Ordovician (Pogonip) sponge fauna. *Proceedings of the United States National Museum* 91, 91–102.

Bengtson, S. 1979: Sponges. *In* Jaanusson, V., Laufeld, S. & Skoglund, R. (eds): *Lower Wenlock Faunal and Floral Dynamics – Vattenfallet Section, Gotland.* Sveriges Geologiska Undersökning C *762*, 61–62.

Bengtson, S. 1981: *Atractosella*, a Silurian alcyonacean octocoral. *Journal of Paleontology 55*, 281–294.

Beresi, M.S. & Rigby, J.K. 1993: The Lower Ordovician sponges of San Juan, Argentina. *Brigham Young University Geology Studies 39*, 1–64.

Billings, E. 1859: Fossils of the Calciferous sandrock, including some of the deposits of White limestone at Mingan, supposed to belong to the formation. *Canadian Naturalist and Geologist and Proceedings of the Natural History Society of Montreal 4*, 345–346.

Billings, E. 1865: On some new or little-known species of lower Silurian fossils from the Potsdam Group (Primordial Zone). *In Palaeozoic Fossils, volume 1. Containing Descriptions and Figures of New or Little Known Species of Organic Remains from the Silurian Rocks.* Geological Survey of Canada, 426 pp.

Bogoyavlenskaya, O.V. 1969: On constructing the classification of the stromatoporoids. *Paleontologičeskij Žurnal 1969*, 12–27. [In Russian].

Borchiellini, C., Chombard, C., Manuel, M., Alivon, E., Vacelet, J. & Boury–Esnault, N. 2004: Molecular phylogeny of Demospongiae: implications for classification and scenarios of character evolution. *Molecular Phylogenetics and Evolution 32*, 823–837.

Botting, J.P. 2004: An exceptional Caradoc sponge fauna from the Llanfawr quarries, central Wales. *Journal of Systematic Palaeontology 2*, 31–63.

Botting, J.P. 2005: Exceptionally well-preserved Middle Ordovician sponges from the Llandegley Rocks, Lagerstätte. *Palaeontology 48*, 577–617.

Botting, J.P. & Butterfield, N.J. 2005: Reconstructing basal sponge relationships using the Burgess Shale fossil *Eiffellia*. *Proceedings of the National Academy of Sciences 102*, 1554–1559.

Botting, J.P. & Rhebergen, F. 2011: A remarkable new Middle Sandbian (Ordovician) hexactinellid sponge in Baltic Erratics. *Scripta Geologica 143*, 1–14.

Boucot, A.J. 1975: *Evolution and Extinction Rate Controls*, 427 pp. Elsevier, New York.

Brett, C.E., Boucot, A.J. & Jones, B. 1993: Absolute depths of Silurian benthic assemblages. *Lethaia 26*, 25–40.

Brückner, A. 2006: Taxonomy and paleoecology of lyssacinosan Hexactinellida from the Upper Cretaceous (Coniacian) of Bornholm, Denmark, in comparison with other Postpaleozoic representatives. *Abhandlungen der Senckenbergischen Naturforschenden Gesellschaft 564*, 1–103.

Brückner, A. & Janussen, D. 2005: *Rossella bromleyi* sp. nov.: the first entirely preserved fossil sponge species of the genus *Rossella* (Hexactinellida) from the Upper Cretaceous of Bornholm, Denmark. *Journal of Paleontology 79*, 21–28.

Calner, M., Jeppsson, L. & Munnecke, A. 2004: The Silurian of Gotland – Part I: Review of the stratigraphic framework, event stratigraphy, and stable carbon and oxygen isotope development. 113–131. *In* Munnecke, A. & Servais, T. (eds): *International Symposium on Early Palaeozoic Palaeogeography and Palaeoclimate*, volume 5. Erlanger Geologische Abhandlungen, Sonderband. 156 pp.

Carrera, M.G. 2006: The new genus *Multispongia* (Porifera) from the Lower Ordovician limestones of the Argentine Precordillera. *Ameghiniana 43*, 1–6.

Dendy, A. 1924: On an orthogenetic series of growth forms in certain tetraclonid sponge spicules. *Proceedings of the Royal Society of London (Section B) 97*, 243–250.

Dohrmann, M., Janussen, D., Reitner, J., Collins, A.G. & Wörheide, G. 2008: Phylogeny and evolution of glass sponges (Porifera, Hexactinellida). *Systematic Biology 57*, 388–405.

Duncan, P.M. 1879: On some spheroidal lithistid Spongida from the Upper Silurian Formation of New Brunswick. *Annals and Magazine of Natural History 5*, 84–91.

Eriksson, K. & Hagenfeldt, S.E. 1997: Acritarch assemblages from the Lower Silurian (Llandovery – Wenlock) in the Grötlingboborrningen 1 core, Gotland, Sweden. *GFF 119*, 13–16.

Erpenbeck, D. & Wörheide, G. 2008: On the molecular phylogeny of sponges (Porifera). *Zootaxa 1668*, 107–126.

Finks, R.M. 1960: Late Paleozoic sponge faunas of the Texas region: the siliceous sponges. *American Museum of Natural History Bulletin 120*, 1–160.

Finks, R.M. 2003: Paleozoic Demospongea: Morphology and Phylogeny. 63–80. *In* Kaesler, R.L. (ed.): *Treatise on Invertebrate Paleontology, Part E, Porifera (Revised) 2: Introduction to the Porifera*, xxvii + 349 pp. The Geological Society of America and The University of Kansas, Boulder and Lawrence, Kansas.

Finks, R.M. & Rigby, J.K. 2004a: Paleozoic Demosponges. 9–173. *In* Kaesler, R.L. (ed.): *Treatise on Invertebrate Paleontology, Part E, Porifera (Revised) 3*, xxxi + 872 pp. The Geological Society of America and the University of Kansas Press, Boulder, Colorado and Lawrence, Kansas.

Finks, R.M. & Rigby, J.K. 2004b: Paleozoic Hexactinellid Sponges. 319–448. *In* Kaesler, R.L. (ed.): *Treatise on Invertebrate Paleontology, Part E, Porifera (Revised), 3*, xxxi + 872 pp. The Geological Society of America and the University of Kansas Press, Boulder, Colorado and Lawrence, Kansas.

Flodén, T. 1980: Seismic stratigraphy and bedrock geology of the Central Baltic. *Stockholm Contributions in Geology 35*, 1–240.

Foerste, A.E. 1916: Notes on the Cincinnatian fossil types. *Bulletin of the Scientific Laboratories of Denison University 18*, 285–355.

Fredén, C. (ed.) 1994. *Geology*, 208 pp. National Atlas of Sweden, Stockholm.

de Freitas, T.A. 1989: Silurian *Archaeoscyphia* from the Canadian Arctic: a case for simplified generic taxonomy in the anthaspidellid lithistids. *Canadian Journal of Earth Sciences 26*, 861–879.

de Freitas, T.A. 1991: Ludlow (Silurian) lithistid and hexactinellid sponges, Cape Phillips Formation, Canadian Arctic. *Canadian Journal of Earth Sciences 28*, 2042–2061.

Gerth, H. 1927: Die Spongien aus dem Perm von Timor. *Jaarboek van het Mijnwezen in Nederlandsch – Indië, Verhandelingen voor 1926, 1*, 93–132.

Goldfuss, A. 1826–1833: *Petrefacta Germaniae oder Abbildungen und Beschreibungen der Petrefacten Deutschlands und der Angrenzenden Länder, Band 1, Heft 1–4*, xxvi + 252 pp. Unter Mitwirkung des Grafen Geoge zu Münster, Düsseldorf.

Grahn, Y. 1995: Lower Silurian chitinozoa and biostratigraphy of subsurface Gotland. *GFF 117*, 57–65.

Grant, R.E. 1836: Animal Kingdom. 107–118. *In* Todd, R.B. (ed.), *The Cyclopaedia of Anatomy and Physiology*, volume 1. Sherwood, Gilbert & Piper: London. 813 pp.

von Hacht, U. 1990: Fossile Spongien von Sylt. 103–143. *In* von Hacht, U. (ed.): *Fossilien von Sylt III*, 327 pp. Inge–Maria von Hacht Verlag und Verlagsbuchhandlung, Hamburg.

von Hacht, U. & Rhebergen, F. 1997: *Caryospongia diadema* von Gotland. *Der Geschiebe–Sammler 30*, 67–77.

Hall, J. & Clarke, J.M. 1899 [for 1898]: A memoir of the Palaeozoic reticulate sponges constituting the family Dictyospongidae. *New York State Museum Memoirs 2*, 350 pp., 70 pl.

le Hérissé, A. 1989: Acritarches et kystes d'algues Prasinophycées du Silurien de Gotland, Suède. *Palaeontographia Italica 76*, 57–302.

Hinde, G.J. 1884: *Catalogue of the Fossil Sponges in the Geological Department of the British Museum (Natural History)*, viii + 248 pp. British Museum (Natural History), London.

Hinde, G.J. 1887–1912: *A Monograph of the British Fossil Sponges, volume 1. Sponges From the Palaeozoic and Jurassic Strata*, xix + 254 pp. Palaeontographical Society, London.

Hinde, G.J. & Holmes, W.M. 1892 [for 1891]: On the sponge-remains in the Lower Tertiary strata near Oamaru, Otago, New Zealand. *Journal of the Linnean Society, Zoology 24*, 177–262.

Hooper, J.N.A. 2002: Family Hemiasterellidae Lendenfeld, 1889. 186–195. *In* Hooper, J.N.A. & van Soest, R.W.M. (eds): *Systema Porifera: A Guide to the Classification of Sponges*, 1764 pp. Kluwer Academic/Plenum Publishers, New York.

Howell, B.F. 1940: A new Silurian sponge from Tennessee. *Bulletin of the Wagner Free Institute of Sciences 15*, 45–48.

Jaanusson, V., Laufeld, S. & Skoglund, R. (eds), 1979: Lower Wenlock faunal and floral dynamics – Vattenfallet section, Gotland. *Sveriges Geologiska Undersökning C 762*, 1–294.

Jeppsson, L. 1998: Silurian oceanic events: summary of general characteristics. *In* Landing, E. & Johnson, M.E. (eds): *Silurian Cycles. Linkages of Dynamic Stratigraphy with Atmospheric, Oceanic and Tectonic Changes.* James Hall Centennial Volume. New York State Museum Bulletin *491*, 239–257.

Jeppsson, L., Viira, V. & Männik, P. 2004: Silurian conodont-based correlations between Gotland (Sweden) and Saarema (Estonia). *Geological Magazine 131*, 201–218.

Jeppsson, L., Eriksson, M. & Calner, M. 2006: A latest Llandovery to latest Ludlow high-resolution biostratigraphy based on the Silurian of Gotland – a summary. *GFF 128*, 109–114.

Jin, J. & Copper, P. 2000: Late Ordovician and Early Silurian pentamerid brachiopods from Anticosti Island, Québec, Canada. *Palaeontographica Canadiana 18*, 1–94.

Johns, R.A. 1994: *Ordovician Lithistid Sponges of the Great Basin*, 199 pp. Nevada Bureau of Mines and Geology. Open-file Report 94-1.

Johnson, M.E., Baarli, B.G., Nestor, H., Rubel, M. & Worsley, D. 1991: Eustatic sea-level patterns from the Lower Silurian (Llandovery Series) of southern Norway and Estonia. *Geological Society of America Bulletin 103*, 315–335.

van Kempen, T.M.G. 1978: Anthaspidellid sponges from the Early Paleozoic of Europe and Australia. *Neues Jahrbuch für Geologie und Paläontologie, Abhandlungen 156*, 305–337.

van Kempen, T.M.G. 1983: Een *Caryospongia diadema* (Klöden) Rauff (spons) van Gotland. *Grondboor en Hamer 37*, 100–104.

van Kempen, T.M.G. 1990a: Two Baltic Ordovician chiastoclonellids (Porifera) from the island of Sylt (NW. Germany). 151–178. *In* von Hacht U. (ed.): *Fossilien von Sylt III*, 327 pp. Inge–Maria von Hacht Verlag und Verlagsbuchhandlung, Hamburg.

van Kempen, T.M.G. 1990b: On the oldest tetraxon megascleres. 9–16. *In* Rützler, K. (ed.): *New Perspectives in Sponge Biology*, 533 pp. Smithsonian Institution Press, Washington, DC.

King, R.H. 1943: New Carboniferous and Permian sponges. *State Geological Survey of Kansas Bulletin 47*, 1–36.

Klöden, K.F. 1834: *Die Versteinerungen der Mark Brandenburg, insonderheit diejenigen, welche sich in den Rollsteinen und Blöcken der südbaltischen Ebene finden*, 381 pp. Lüderitz, Berlin.

Kühn, O. 1927: Zur Systematik und Nomenklatur der Stromatoporen. *Zentralblatt Mineralogie, Geologie und Paläontologie B 1927*, 546–551.

Kühn, O. 1939: Eine neue Familie der Stromatoporen. *Zentralblatt Mineralogie, Geologie und Paläontologie B 1939*, 338–345.

de Laubenfels, M.W. 1955: Sponges. 21–112. *In* Moore, R.C. (ed.): *Treatise on Invertebrate Paleontology, Part E, Archaeocyatha and Porifera*, xviii + 122 pp. The Geological Society of America and the University of Kansas Press, New York and Lawrence.

Laufeld, S. & Jeppsson, L. 1976: Silicifications and bentonites in the Silurian of Gotland. *Geologiska Föreningens I Stockholm Förhandlingar 98*, 31–44.

Lévi, C. 1953: Sur une nouvelle classification des Démosponges. *Comptes Rendus de l' Academie des Sciences de Paris 236*, 853–855.

Lindström, G. 1885. *List of the Fossils of the Upper Silurian Formation of Gotland*, 20 pp. Kongl, Boktryckeriet, Stockholm.

Lindström, G. 1888a: *List of the Fossil Faunas of Sweden. I. Cambrian and Lower Silurian*, 24 pp. Palaeontological Department of the Swedish State Museum (Natural History), Stockholm.

Lindström, G. 1888b: *List of the Fossil Faunas of Sweden. II. Upper Silurian*, 29 pp. Palaeontological Department of the Swedish State Museum (Natural History), Stockholm.

Lindström, G. 1888c: Ueber die Schichtenfolge des Silur auf der Insel Gotland. *Neues Jahrbuch für Mineralogie, Geologie und Paläontologie 1*, 145–164.

M'Coy, F. 1850: On some new genera and species of Silurian Radiata in the collection of the University of Cambridge. *Annals and Magazine of Natural History (Series 2), 6*, 270–290.

Manten, A.A. 1971: *Silurian Reefs of Gotland. Developments in Sedimentology 13*, 539 pp. Elsevier Publishing Company, Amsterdam.

Marshall, W. 1876: Ideen über die Verwandtschaftsverhältnisse der Hexactinelliden. *Zeitschrift für Wissenschaftliche Zoologie 27*, 113–136.

Martinsson, A. 1967: The succession and correlation of ostracode faunas in the Silurian of Gotland. *Geologiska Föreningens i Stockholm Förhandlingar 89*, 350–386.

Mehl-Janussen, D. 1999: Die frühe Evolution der Porifera. Phylogenie und Evolutionsökologie der Poriferen im Paläozoikum mit Schwerpunkt der desmentragenden Demospongiae ("Lithistide"). *Münchner Geowissenschaftlichen Abhandlungen. Reihe A. Geologie und Paläontologie 37*, 1–72.

Miller, S.A. 1889: Class Porifera. 152–167. *In North American Geology and Palaeontology*. Published by the author, Cincinatti.

Mori, K. 1969: Stromatoporoids from the Silurian of Gotland, Part I. *Stockholm Contributions in Geology 19*, 1–100.

Muir, L.A., Botting, J.P., Carrera, M. & Beresi, M. 2013: Cambrian, Ordovician and Silurian non-stromatoporoid Porifera. *In* Harper, D.A.T. & Servais, T. (eds): *Early Palaeozoic Palaeobiogeography and Palaeogeography*. Geological Society, London, Memoirs *38*, 81–95.

Narbonne, G.M. & Dixon, O.A. 1984: Upper Silurian lithistid sponge reefs on Somerset Island, Arctic Canada. *Sedimentology 31*, 25–40.

Nestor, H. 1964: *Ordovician and Llandoverian Stromatoporoidea of Estonia*, 112 pp. Akademiâ Nauk Estonskoj SSR, Institut Geologii, Tallinn. [In Russian with English summary].

Nestor, H. 1966: *Wenlockian and Ludlovian Stromatoporoidea of Estonia*, 87 pp. Akademiâ Nauk Estonskoj SSR, Institut Geologii, Tallinn. [In Russian with English summary].

Nicholson, H.A. & Murie, J. 1878: On the minute structure of *Stromatopora* and its allies. *Linnean Society, Journal of Zoology 14*, 187–246.

Oswald, F. 1847: Ueber die Petrefacten von Sadewitz. *Uebersicht der Arbeiten und Veränderungen der Schlesischen Gesellschaft für vaterländische Kultur im Jahre 1846 (Breslau)*, 56–65.

Pickett, J. 1969: Middle and Upper Palaeozoic sponges from New South Wales. *Memoirs of the Geological Survey of New South Wales, Palaeontology 16*, 1–24.

Pisera, A. 2002: †Fossil Lithistids: An overview. 388–402. *In* Hooper, J.N.A. & van Soest, R.W.M. (eds): *Systema Porifera: A Guide to the Classification of Sponges*, 1764 pp. Kluwer Academic/Plenum Publishers, New York.

Quenstedt, F.A. 1878: *Petrefactenkunde Deutschlands. 5. Korallen (Schwämme)*, 558 pp. Fues's Verlag, Leipzig.

Rauff, H. 1893: Palaeospongiologie. Erster oder allgemeiner Theil und Zweiter Theil, erste Hälfte. *Palaeontographica 40*, 1–232.

Rauff, H. 1894: Palaeospongiologie. Erster oder allgemeiner Theil und Zweiter Theil, erste Hälfte. *Palaeontographica 41*, 233–346.

Rauff, H. 1895: Palaeospongiologie. Zweiter Theil. Fortsetzung. Spongien des Silurs. *Palaeontographica 43*, 223–272, 347–395.

Reid, R.E.H. 1963: A classification of the Demospongia. *Neues Jahrbuch für Geologie und Paläontologie, Monatshefte 4*, 196–207.

Reid, R.E.H. 1968: *Hyalostelia smithii* (Young & Young) and the sponge genus *Hyalostelia* Zittel (Class Hexactinellida). *Journal of Palaeontology 42*, 243–248.

Reiswig, H.M. 2002. Order Amphidiscosida Schrammen, 1924. 1231. *In* Hooper, J.N.A. & van Soest, R.W.M. (eds): *Systema Porifera: A Guide to the Classification of Sponges*, 1764 pp. Kluwer Academic/Plenum Publishers, New York.

Reiswig, H.M. & Wheeler, B. 2002. Family Euretidae Zittel, 1877. 1301–1331. *In* Hooper, J.N.A. & van Soest, R.W.M. (eds): *Systema Porifera: A Guide to the Classification of Sponges*, 1764 pp. Kluwer Academic/Plenum Publishers, New York.

Rhebergen, F. 1997: Twee nieuwe sponzensoorten als zwerfsteen in Nederland: *Chiastoclonella* sp. en *Syltispongia ingemariae*. *Grondboor en Hamer 51*, 138–141.

Rhebergen, F. 2004: A new Ordovician astylospongiid sponge (Porifera) as an erratic from Baltica. *Netherlands Journal of Geosciences 83*, 255–265.

Rhebergen, F. 2005: Sponges (Porifera) from Silurian strata on Gotland, Sweden. *GFF 27*, 211–216.

Rhebergen, F. 2007: Baltic Ordovician compound sponges as erratics on Gotland (Sweden), in northern Germany and the eastern Netherlands. *Netherlands Journal of Geosciences 86*, 365–378.

Rhebergen, F. & von Hacht, U. 2000: Ordovician erratic sponges from Gotland, Sweden. *GFF 122*, 339–349.

Rhebergen, F. & van Kempen, T.M.G. 2002: An unusual Silurian erratic astylospongiid (Porifera) from Gotland, Sweden. *GFF 124*, 185–192.

Rhebergen, F., Eggink, R.G., Koops, T. & Rhebergen, B. 2001: Ordovicische zwerfsteensponzen. Staringia 9. *Grondboor en Hamer 55*, 1–144.

Rigby, J.K. 1973: A new anthaspidellid sponge from the Silurian of Lake Timiskaming, Quebec. *Journal of Palaeontology 47*, 801–804.

Rigby, J.K. 1977: Two new Middle Ordovician sponges from Foxe Plain, southeastern District of Franklin. *Geological Survey of Canada Bulletin 269*, 121–129.

Rigby, J.K. 1986: Late Devonian sponges from Western Australia. Geological Survey of Western Australia, Report 18, vii + 59.

Rigby, J.K. 2004a: Case 3316. Hindia Duncan, 1879 (Porifera): proposed conservation. *Bulletin of Zoological Nomenclature 61*, 80–82.

Rigby, J.K. 2004b: Classification. 1–8. *In* Kaesler, R.L. (ed.): *Treatise on Invertebrate Paleontology, Part E, Porifera (Revised) 3*, xxxi + 872 pp. The Geological Society of America and the University of Kansas, Boulder, Colorado and Lawrence, Kansas.

Rigby, J.K. & Bayer, T.N. 1971: Sponges of the Ordovician Maquoketa Formation in Minnesota and Iowa. *Journal of Palaeontology 45*, 608–627.

Rigby, J.K. & Chatterton, B.D.E. 1989: Middle Silurian Ludlovian and Wenlockian Sponges from Baillie–Hamilton and Cornwallis Islands, Arctic Canada. *Geological Survey of Canada 391*, 1–37.

Rigby, J.K. & Chatterton, B.D.E. 1994: New middle Silurian hexactinellid sponge from the Mackenzie Mountains, Northwest Territories, Canada. *Journal of Paleontology 68*, 218–223.

Rigby, J.K. & Chatterton, B.D.E. 1999: Silurian (Wenlock) demosponges from the Avalanche Lake area of the Mackenzie Mountains, southwestern District of Mackenzie, Northwest Territories, Canada. *Palaeontographica Canadiana 16*, 1–39.

Rigby, J.K. & Chidsey, T.C. 1976: A well-preserved *Calycocoelia typicalis* Bassler (Porifera) from the Ordovician Fort Peña Formation of western Texas. *Brigham Young University Geology Studies 23*, 3–8.

Rigby, J.K. & Clement, C.R. 1995: Demosponges and hexactinellid sponges from the Lower Devonian Ross Formation of west-central Tennessee. *Journal of Palaeontology 69*, 211–232.

Rigby, J.K. & Collins, D. 2004: Sponges of the Middle Cambrian Burgess Shale and Stephen formations, British Columbia. *ROM Contributions in Science 1*, 1–155.

Rigby, J.K. & Desrochers, A. 1995: Lower and Middle Ordovician Demosponges of the Mingan Islands, Gulf of St. Lawrence, Quebec. *Palaeontological Society Memoir 41*, 1–35.

Rigby, J.K. & Dixon, O.A. 1979: Sponge fauna of the Upper Silurian Red Bay Formation, Somerset Island, District of Franklin, Arctic Canada. *Journal of Paleontology 53*, 587–627.

Rigby, J.K. & Lenz, A.C. 1978: A new Silurian astylospongid sponge from Baillie–Hamilton Island, Canadian Arctic Archipelago. *Canadian Journal of Earth Sciences 15*, 157–162.

Rigby, J.K. & Pollard-Bryant, T.L. 1979: Fossil sponges of the Mississippian Fort Payne Chert in northeastern Alabama. *Journal of Paleontology 53*, 1005–1012.

Rigby, J.K. & Webby, B.D. 1988: Late Ordovician sponges from the Malongulli Formation of central New South Wales, Australia. *Palaeontographica Americana 56*, 1–147.

Rigby, J.K., Pisera, A., Wrzolek, T.T. & Racki, G. 2001: Upper Devonian sponges from the Holy Cross Mountains, Central Poland. *Palaeontology 44*, 447–488.

Roemer, F. 1848: Über eine neue Art der Gattung *Blumenbachium* (König) und mehrere unzweifelhafte Spongien in obersilurischen Kalkschichten der Grafschaft Decatur im Staate Tennessee in Nord-Amerika. *Neues Jahrbuch für Mineralogie, Geologie und Paläontologie 1848*, 680–686.

Roemer, F. 1860. *Silurische Fauna des westlichen Tennessee. Eine Paläontologische Monographie*, 97 pp. Edvard Trewendt Verlag, Breslau.

Rukhin, L.B. 1938: Lower Paleozoic corals and stromatoporoids of the upper reaches of the Kolyma River basin. [Contribution to the knowledge of the Kolyma–Indigirka land. Series 2.]. *Geology and Geomorphology 10*, 1–119. [In Russian].

Salter, J.W. 1861: Descriptions and lists of fossils. Appendix. 132–151. *In Geology of the Neighbourhood of Edinburgh*. Memoirs of the Geological Survey of Great Britain, London, Sheet 32.

Sapelnikov, V.P. 1976: *Reveroides – novyy siluriyskiy rod pentamerid s vostochnogo sklona Urala*. Akademiya Nauk SSSR, Uralskiy Nauchnyy Tsentr, Instituta Geologii i Geokhimii, Ezhegodnik, 1975, 3. [In Russian].

Sarà, M. 2002: Family Tethyidae Gray, 1848. 245–265. *In* Hooper, J.N.A. & van Soest, R.W.M. (eds): *Systema Porifera: A Guide to the Classification of Sponges*, 1764 pp. Kluwer Academic/Plenum Publishers, New York.

Schlüter, C. 1884: Über *Astylospongia Gothlandica* sp.n. *Verhandlungen des Naturhistorischen Vereines der Preussischen Rheinlande und Westfalens, 41–5*, 79–80.

Schmidt, O. 1870: *Grundzüge Einer Spongien–Fauna des Atlantischen Gebietes*, iv + 88 pp. W. Engelmann, Leipzig.

Schmidt, F. 1891: Einige Bemerkungen über das Baltische Obersilur in Veranlassung der Arbeit des Prof. W. Dames über die Schichtenfolge der Silurbildungen Gotlands. *Bulletin de l'académie Impériale des Sciences de St.-Pétersbourg 1*, 121–139.

Schrammen, A. 1910: Die Kieselspongien der oberen Kreide von Nordwestdeutschland, 1. Theil, Tetraxonia, Monaxonia und Silicea incert. sedis. *Palaeontographica, Supplement 5*, 1–175.

Schuchert, C. & Cooper, G.A. 1931: Synopsis of the brachiopod genera of the suborders Orthoidea and Pentameroidea, with notes on the Telotremata. *American Journal of Science 20*, 241–251.

Schulz, E. 1883: Die Eifelkalkmulde von Hillesheim, Nebst einem palaeontologischen Anhang. *Königliche Preussische Geologische Landesanstalt und Bergakademie zu Berlin, Jahrbuch für 1882*, 158–250.

Seamon, D. & Zajonc, A. (eds) 1998. *Goethe's Way of Science: A Phenomenology of Nature*, 324 pp. State University of New York Press, New York.

van Soest, R.W.M. 2002: Family Solassellidae Lendenfeld, 1887. 279–280. *In* Hooper, J.N.A. & van Soest, R.W.M. (eds): *Systema Porifera: A Guide to the Classification of Sponges*, 1764 pp. Kluwer Academic/Plenum Publishers, New York.

Sollas, W.J. 1875: Sponges. 427–446. *Encyclopedia Brittanica*. Encyclopedia Brittanica Co., Edinburgh.

Spjeldnaes, N. 1976: An anomalous occurrence of the Silurian Visby Marl on Gotland, Sweden. *Geologiska Föreningens i Stockholm Förhandlingar 98*, 367–370.

Stearn, C.W. 1980: Classification of the Paleozoic stromatoporoids. *Journal of Paleontology 54*, 881–902.

Stearn, C.W., Webby, B.D., Nestor, H. & Stock, C.W. 1999: Revised classification and terminology of Palaeozoic stromatoporoids. *Acta Palaeontologica Polonica 44*, 1–70.

Stolley, E. 1900: Geologische Mittheilungen von der Insel Sylt. II. Zur Geologie der Insel Sylt. II. Cambrische und silurische Gerölle im Miocän. *Archiv für Anthropologie und Geologie Schleswig–Holsteins IV, 1*, 3–49.

Stolley, E. 1929: Geologica varia von den Nordseeinseln. Sonderabdruck aus dem 23. *Jahresbericht des Niedersächsischen*

Geologischen Vereins (Geologische Abteilung der Naturhistorischen Gesellschaft zu Hannover), 31–111.

Stuckenberg, A. 1895: Korallen und Bryozoen der Steinkohlenablagerungen des Ural und des Timan. *Mémoires du Comité Géologique, St. Petersbourg 10*, 244 pp.

Svantesson, S.-I. 1976: Granulometric and petrographic studies of till in the Cambro–Silurian Area of Gotland, Sweden, and studies of the ice recession in northern Gotland. *Striae 2*, 1–80.

Tabachnik, K.R. 2002: Family Rossellidae Schulze, 1885. 1441–1505. *In* Hooper, J.N.A. & van Soest, R.W.M. (eds): *Systema Porifera: A Guide to the Classification of Sponges*, 1764 pp. Kluwer Academic/Plenum Publishers, New York.

Topsent, É. 1898: Introduction à l'étude monographique des Monaxonides de France, Classification des Hadromerina. *Archives de Zoologie Expérimentale et Générale 4*, 91–113.

Ulrich, E.O. 1889: Preliminary description of new lower Silurian sponges. *The American Geologist 3*, 233–248.

Ulrich, E.O. 1890: American Paleozoic sponges. *Illinois Geological Survey. (Palaeontology of Illinois, 2, 3). Bulletin 8*, 209–241.

Ulrich, E.O. & Everett, O. 1889: Lower Silurian Sponges. *Illinois Geological Survey (Palaeontology of Illinois, 2, 5). Bulletin 8*, 255–282.

Walcott, C.D. 1886: Second contribution to the studies of the Cambrian faunas of North America. *United States Geological Survey Bulletin 30*, 1–369.

Webby, B.D. 1993: Evolutionary history of Palaeozoic Labechiida (Stromatoporoidea). *Memoir of the Association of Australian Palaeontologists 15*, 57–67.

Wiman, C. 1907: Studien über das Nordbaltische Silurgebiet. II. *Bulletin of the Geological Institute of the University of Uppsala 8*, 74–168.

Young, J. & Young, J. 1877: On the Carboniferous *Hyalonema* and other sponges from Ayrshire. *Annals and Magazine of Natural History (Series 4), 20*, 425–435.

Zhan-Qiu, D. 1981: Upper Permian sponges from Laibin of Guangxi. *Acta Palaeontologica Sinica 20*, 418–427. [In Chinese with English summary].

Zhan-Qiu, D. 1990: Sponges and receptaculitids from Ningqiang Formation (Late Llandovery) of Guanyuan Sichuan. *Acta Palaeontologica Sinica 29*, 581–591. [In Chinese, English summary].

von Zittel, K.A. 1877: Beiträge zur Systematik der Fossilen Spongien. *Neues Jahrbuch für Mineralogie, Geologie und Paläontologie 1877*, 337–378.

von Zittel, K.A. 1878a: Studien über fossile Spongien, 2. Lithistidae. A. Allgemeiner Theil. *Abhandlungen der Mathematisch–Physikalischen Classe der Königlich Bayerischen Akademie der Wissenschaften 13*, 65–154.

von Zittel, K.A. 1878b: Studien über fossile Spongien, Dritte Abtheilung: Monactinellidae, Tetractinellidae und Calcispongiae. *Abhandlungen der Mathematisch–Physikalischen Classe der Königlich Bayerischen Akademie der Wissenschaften 13*, 93–138.

von Zittel, K.A. 1895: *Grundzüge der Paläontologie (Paläozoologie), Abteiling 1, Invertebrata*, viii + 971 pp. R. Oldenbourg, München and Leipzig.